HABITABLE PLANETS FOR MAN

STEPHEN H. DOLE

Habitable Planets for Man was originally published by Blaisdell Publishing Company in 1964. This RAND edition reflects the original layout.

Library of Congress Cataloging-in-Publication Data

978-0-8330-4227-9

Cover Design by Peter Soriano

Published 2007 by the RAND Corporation
1776 Main Street, P.O. Box 2138, Santa Monica, CA 90407-2138
1200 South Hayes Street, Arlington, VA 22202-5050
4570 Fifth Avenue, Suite 600, Pittsburgh, PA 15213-2665
RAND URL: http://www.rand.org/
To order RAND documents or to obtain additional information, contact
Distribution Services: Telephone: (310) 451-7002;
Fax: (310) 451-6915; Email: order@rand.org

To Minda

PREFACE

The space age is still very much in its infancy. To attempt at this early date to predict the ultimate future of space flight and its impact on human affairs would be like trying to forecast the complete career of a child barely out of his cradle. Many unforeseen events can occur, and many surprises are undoubtedly in store for us. But, as one who has confidence in man's problem-solving ingenuity, the author believes that space flight will eventually come of age and will live up to the promise implied in the term "astronautics" (*astro-*, star; *-nautics*, pertaining to travel). Where will space flight lead us, as man gradually develops his ability to tap ever more powerful sources of energy and attain higher and higher velocities through space?

This book is concerned with our ultimate destinations in space rather than with methods of propulsion or the technical problems of getting from here to there. The technical questions of traveling across interstellar distances have been treated at length elsewhere. Our present concern is, Where will man eventually want to go and what will he find when he gets there?

Answers to many of the questions raised here have never before been attempted; consequently, there is no established consensus to which appeal can be made. In many cases, completely new and unorthodox approaches had to be worked out in order to arrive at suitable solutions (for example, the concepts of general planetology).

Planets and stars are supremely important to every member of the human race. Although most people never give it much thought, we live on a planet and are warmed and nourished by a star—not just any planet and not just any star, but a particular kind of planet with certain essential characteristics, and a particular kind of star. The central purpose of this book is to spell out the necessary requirements of planets on which human beings as a biological species (*Homo sapiens*) can live, and the essential properties required of the stars that provide heat and light to such planets.

To do justice to the many broad fields of science involved in this theme, the author should properly be an authority in several dozen disciplines, ranging from aerodynamics to zoology; but clearly no one person can fully master more than a fraction of all that is relevant in even one of these fields. In full recognition of this, the author has attempted to avoid the compartmentalization of science and has tried to bring to bear on the subject as much pertinent information as he could. Any errors of understanding or interpretation are completely his own.

This study was prepared as part of the continuing program of research undertaken by The RAND Corporation for the United States Air Force.

STEPHEN H. DOLE
Santa Monica, California
November, 1962

CONTENTS

LIST OF FIGURES

LIST OF TABLES

THE BRIGHTEST STAR IN CASSIOPEIA

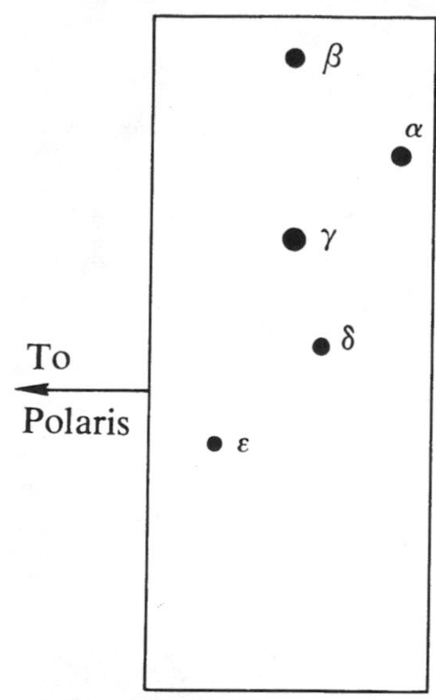

As seen from the Earth

Alpha Centauri, our nearest neighbor in space beyond the solar system and one of the brightest stars of the Southern Hemisphere, is actually a triple star system. The two larger stars, A and B, revolve about each other in a period of 80 years, at a mean separation of about 23 astronomical units [an astronomical unit (A.U.) is the mean distance between the Sun and the Earth]; the third member, Alpha Centauri C, some 10,000 astronomical units removed, revolves about the pair A, B in a period of the order of a million years. Either (or both) of the stars A and B *could* possess a habitable planet; C could not.

A hypothetical intelligent inhabitant of such a planet, looking out at his night sky, would see almost identically the same familiar constellations we see from the Earth: the Big Dipper, Orion, and all the others. But there would be an outstanding addition. Instead of five bright stars in Cassiopeia, he would see six. The sixth and brightest (of visual magnitude zero), extending the zigzag pattern of stars one more step, would be our Sun (see Figure 1).

As far as our hypothetical intelligent observer would be able to discern, there would be nothing intrinsically remarkable about this nearby bright star; it would look very much the same as many others within his view. He would have no way of detecting the fact that it actually possessed a life-bearing planet in orbit about it.

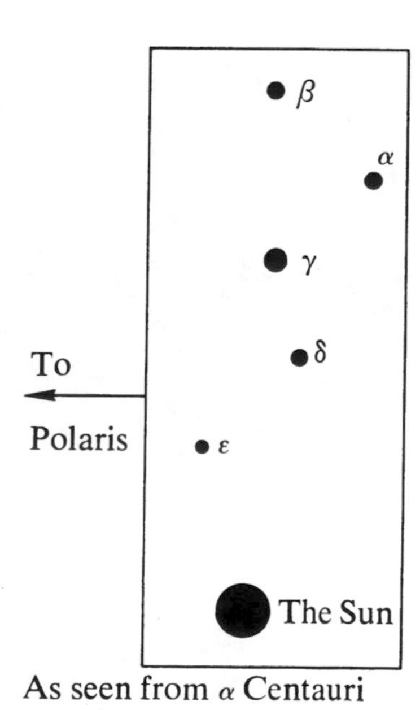

As seen from α Centauri

Figure 1—The principal stars in Cassiopeia

CHAPTER 1

Introduction

THE PROMISE OF SPACE

One of the long-term goals of space flight, and possibly the most important, will be to find planets of other stars on which human beings could settle and live as freely and comfortably as they do on the Earth's surface. This book comprises an attempt to make a quantitative estimate of the probabilities of finding such planets, where they might be found, and how many habitable planets there may be in our Galaxy.

In order to be called habitable, for our purposes, a planet must provide an acceptable environment for human beings. The procedure will be as follows: First, the human environmental requirements are spelled out; next, the general properties of the massive bodies that we call planets are examined in detail; and then, by analysis of these factors together, a class of planets is described that possesses the specific physical and astronomical parameters compatible with the requirements of human beings. Finally, the probabilities are estimated that a given main-sequence star, as a function of its mass, will have at least one habitable planet in orbit about it. On this basis, the stars nearest to the Earth that are most likely to possess habitable planets are selected and listed.

Other related subjects discussed here include descriptions of the special and unusual kinds of habitable planets that may eventually be discovered, the effects that new planetary environments might have on people who migrate to these planets, procedures for searching for habitable planets, and an appraisal of the Earth as a planet and how its habitability would be changed if some of its basic properties were altered.

It will be appreciated that many of the ideas presented in this book can not be proved (nor can they be disproved); rather, they are based on inferences and reasonable assumptions within the framework of our present knowledge.

At the present writing no human being has ever traveled more than a few hundred miles from the thin spherical film that we call our world. The entire human species before us has lived and died on the surface of a sphere less than 4000 miles in radius, a sphere that revolves about a fairly ordinary star in one of the spiral arms of an island galaxy known as the Milky Way.* Yet who can doubt that within our lifetimes men will travel to the Moon and probably to other planets of the solar system?

The human race is about to make its first expedition into space. Why are some men eager to leave the comforts of their homes to risk their lives in the inhospitable reaches of this vast unknown? It is as easy to understand and as difficult to explain as the migrations of the ancestors of the Tierra del Fuegans and the Hawaiian Islanders, as the voyages of Columbus, as the perseverance of Admiral Byrd, as the determination of Hillary. It is just those qualities that make us human beings which challenge men to penetrate the unknown, to push back the frontiers of knowledge, to become explorers and pioneers.

The plans now being made for manned exploration of the space near the Earth may be likened to the first efforts of a baby who is just beginning to creep. It is a learning process with only short-term goals, hard to justify in terms of the short view. Our first, faltering steps will take us to the Moon. More confident steps will lead to exploring the surface of Mars and the cloud cover of Venus and, later on, the other planets of the solar system. But where will the process take us when we have really learned how to traverse space with comparative assurance and ease?

This book assumes the long-term viewpoint: it looks forward to a time when man will be able to seek and find habitable planets beyond our solar system. We can not do it now, nor do we really know how it can be done within the scope of our present limited knowledge, although the possibilities of space flights over interstellar distances have been discussed by a few writers (see, for example, Sänger, 1956, 1961; Spencer and Jaffe, 1962; and Bussard, 1960). We are barely started, however, on our upward climb as self-aware and knowledge-seeking creatures. There is much more to learn than has already been learned—and the human race has a long future before it. We are concerned here with *destinations* rather than with the means of getting to them. In view of the rapid and accelerating strides of science and technology during the past 50 or 60 years, it would

---

* Our Galaxy contains more than a hundred billion stars; there are billions of other galaxies in the universe.

take an extreme pessimist indeed to say with conviction, "Man will *never* learn how to travel through space fast enough to reach the stars in one lifetime or even several lifetimes."

What are some of the long-term goals of manned space flight? Surely some of the major ones are to increase our knowledge and understanding of the universe; to attempt to learn about the origin of matter, the origin of galaxies, and the origin of life; and, possibly, to find new sources of energy and raw materials. But there is another: to find new habitable planets.

As it is now, our planet Earth supports the entire human race. It is possible that one catastrophe could completely destroy our form of life. But if the human race were living on a number of planets scattered around the Galaxy, its immortality would be virtually assured. One of the most important facets of life on the Earth today is the rapidly increasing number of human beings. As population pressures increase, there will be ever-increasing incentives for pioneers to migrate to the next frontier. Who can predict when a scientific discovery or a series of discoveries may completely alter our capabilities for traveling across interstellar distances and seeking out new planets on which to live? Given such capabilities, where would we begin to look for habitable planets? What is the probability of finding them? This book tries to provide some of the answers.

PHILOSOPHY OF APPROACH

How do we commence looking for other habitable planets in the local Galaxy? Objects of planetary size at the distances of the closest stars can not be detected optically, even under the most favorable conditions with the use of the Palomar Observatory's 200-inch telescope, the world's largest "eye" on space.

The present approach will be as follows: wherever possible, use will be made of available factual material and known relationships; where gaps exist in our knowledge, reasonable inferences will be based on models constructed from observations of conditions in the solar system. Some dependence on theory will be necessary where no direct data exist; all known physical and chemical laws must be obeyed elsewhere in the universe.

In the past, many serious writers have speculated about the existence of life on other planets—Percival Lowell (1908), H. Spencer Jones (1949), Harlow Shapley (1959), and Hubertus Strughold (1955), to name a few. But, generally speaking, the subject has been covered in only a qualitative manner. It is the present objective to try to establish reasonable quantitative limits from which the prevalence of habitable planets can be estimated. In setting such an objective, it is necessary to define more precisely what is meant by a habitable planet. Several different categories of habitable planets can be imagined: planets that could support some unknown form of life; planets that could support microscopic forms of carbon-based life, which is life as we know it; and planets that could support various extreme forms of terrestrial life. However, in the present context, a habitable planet is defined as one suitable for human life. This sets aside any speculations about organisms living in seas of liquid ammonia or breathing gaseous sulfur or existing under any other such alien conditions. Further, the use of the term "habitable planet" is meant to imply a planet with surface conditions naturally suitable for human beings, that is, one that does not require extensive feats of engineering to remodel its atmosphere or its surface so that people in large numbers can live there.

Under the present plan we will first circumscribe the environmental conditions required by human beings. Then we will attempt to delineate the astronomical circumstances that produce these requisite environmental conditions. Finally, we will make an estimate of the probabilities that these astronomical concatenations will be found elsewhere in the Galaxy and where they might be found in the immediate neighborhood of the Sun.

It will be realized that many of the questions raised here can not be answered definitively at the present time because of deficiencies in our knowledge of the universe. Therefore many of the conclusions made here must be tentative—based on premises and assumptions that, although believed to be reasonable within our present state of knowledge, may eventually be proved incorrect.

In the spirit of this study, then, conclusions based on clearly labeled assumptions will sometimes be drawn. One of the methods of approach used most frequently will be based on the conviction (based on numerous bits of evidence) that our solar system is not an extraordinarily rare assemblage of bodies, but is rather a typical planetary system, and that its members can be treated as a good, although not numerically large, sample of the types of bodies that exist in proximity to other stars. Every effort will be made to supply a reasonable answer to the following question: In full recognition of the incompleteness of our knowledge, what can now be said regarding the prevalence of habitable planets in our Galaxy?

THE SOLAR SYSTEM

As far as the other planets of our solar system are concerned, none appears to be at all suitable as a habitable planet for human beings under the present definition of the term. No doubt, temporary or even permanent encapsulated colonies will some day be established on the Moon and on Mars, but these will probably never be entirely independent of supplies brought from the Earth.

What is wrong with the other planets of our solar system? Why can they not be considered habitable? Although answers to these questions will be considered later in detail, we list here the conclusions briefly.

Mercury is too small to retain a breathable atmosphere. It is so close to the Sun that its rotation apparently has been stopped by the Sun. Its surface no longer has an alternation of day and night; its sunlit side is too hot, and its dark side, too cold.

Venus, although of the right size (mass), is close enough to the Sun that it has probably lost most of the water produced by volcanism throughout its previous history. As a result, its atmosphere probably contains too much carbon dioxide to be compatible with human life and its surface temperatures are probably too high.

Our satellite, the *Moon*, although at the correct distance from the Sun for habitability, is too small to retain an atmosphere or even to produce one compatible with habitability. Because its rotation is slow, its days are too hot, and its nights are too cold.

Mars is too small to produce or retain an atmosphere suitable for human beings. It is so far from the Sun that, even if it were sufficiently massive, its average surface temperatures would probably still be too low for it to be considered habitable.

Of the remaining planets, *Jupiter*, *Saturn*, *Uranus*, and *Neptune* are so massive that they have retained enormously thick atmospheres, consisting mainly of hydrogen and helium. In addition, they are much too far from the Sun. The satellites of the major planets are too small and too cold. *Pluto* is also too cold.

Of all the planets of the solar system, Mars comes closest to fitting our definition of a habitable planet. It could not, however, support a completely self-sufficient human colony, although it is quite probable that water could be obtained from rocks on the surface, food could be grown locally, and people could live in hermetically sealed "hot houses" having breathable atmospheres. If there is any indigenous life on Mars, it is probably very small, or even microscopic, because of the extreme scarcity of water on the Martian surface.

Since there are no promising possibilities for habitable planets in our solar system, we must search elsewhere—for planets revolving about other stars.

CHAPTER 2

Human Requirements

WHAT IS HABITABLE?

For present purposes, we shall enlarge on our definition of a habitable planet (a planet on which large numbers of people could live without needing excessive protection from the natural environment) to mean that the human population must be able to live there without dependence on materials brought from other planets. In other words, a planet that is habitable can supply all of the physical requirements of human beings and provide an environment in which people can live comfortably and enjoyably. At present, the Earth is the only such planet that we know. Yet it is reasonable to suppose that there may be many other planets in the universe that have the proper combination of surface environmental conditions and on which human beings could live as pleasantly and as freely as they do on the Earth. It is one objective of this book to specify in a quantitative manner, insofar as possible, precisely what astronomical conditions must be met for a planet to be called "habitable." First, let us itemize the specific requirements necessary for planets to be habitable for man.

TEMPERATURE

While it is true that human beings can endure brief periods of extreme heat and cold by using various kinds of protective clothing and other insulation, it is also true that there is a range of temperature that human beings prefer for their everyday existence. This is demonstrated in

Figure 2, which shows the percentages of the Earth's population living in various regions of the Earth's surface as a function of the local mean annual temperature. This illustration shows that human beings prefer to live in regions where the mean annual temperature lies between 40°F and 80°F; essentially, the entire world's population lives in regions where the mean annual temperature is between 32°F and 86°F. It is, of course,

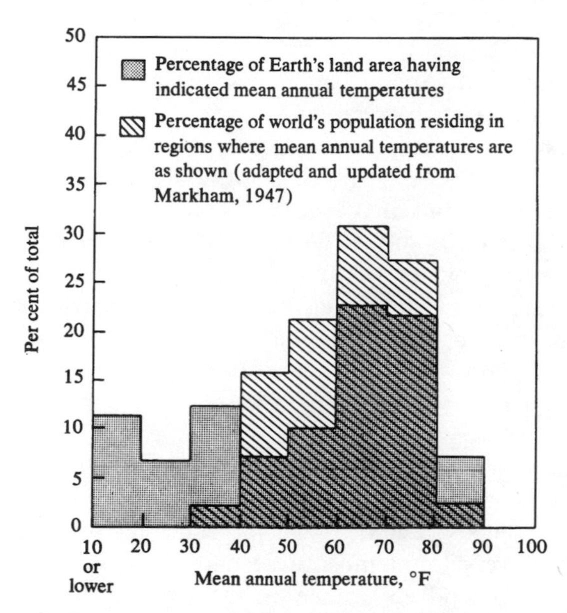

Figure 2. Temperature preferences of the world's population.

not only the human desire for comfort that dictates such a narrow temperature range, but also that these mean temperatures are best tolerated by the agricultural crops and domesticated animals on which man depends for food.

Temperature is obviously not the only parameter that governs where people want to live or do live. Such considerations as water supply, humidity, precipitation patterns, soil conditions, the availability of energy supplies, harbors, altitude, topographical roughness, political pressures, and proximity to other people are also highly important. But temperature is the overriding factor.

There are many instances of living organisms that are tolerant to high temperatures and of other organisms that are tolerant to low temperatures. Some examples follow (Spector, 1956): Certain cyanophyta (blue-green algae), notably *Oscillatoria filiformis*, have been observed living in water at a temperature of 185°F. Another cyanophyte, *Mastigolocladus laminosus*, encounters a maximum of 149°F in its natural environment and is said to tolerate a minimum of −2°F. Ducks have survived 16 days at −40°F air temperature. Cattle have survived 2 weeks at 9°F and, in other experiments, 24 hours at 106°F. *Nadrix sipedon*, a water snake, is reported as having a temperature tolerance range from 32°F to 110°F. The American cockroach (*Periplaneta Americana*) has survived for 24 hours at 120°F. The arctic fir (*Abies excelsa*) is said to carry on photosynthesis from −40°F to +86°F, while the reindeer lichen (*Cladonia rangiferina*) is reported as having an active photosynthesis range from −4°F to +100°F. The morning-glory has been observed in active respiration from 50°F to 140°F.

Other examples of extreme temperature tolerance are as follows (Troitskaia, 1952): The majority of plants cease their functions at temperatures below 32°F but are capable of surviving through temperatures as low as −60°F or −80°F. The seeds of some plants can withstand temperatures even down to −310°F.

Many more examples could be cited to show the remarkable adaptability of some forms of plant and animal life. Nevertheless, the vast majority of important food crops require temperatures within the range of 50°F to 86°F during their growing seasons. The seeds of most herbaceous and woody plants require a temperature within the range of 50°F to 86°F for seed germination. Most fish cannot survive water temperatures much below 32°F or above 86°F. In summary, although a few hardy species of plants and animals can withstand more or less prolonged exposures to very high or very low temperatures and a few species have become adapted to hot or cold environments, most of the plants and animals important to man as sources of food and as suppliers of oxygen through photosynthesis require temperatures above the freezing point of water and below 86°F for survival and active growth.

In addition to the mean-annual-temperature limitation, there is also a limitation because of the daily temperature extremes experienced at the warmest and coldest seasons of the year. It is more difficult to set sharp limits on daily temperature extremes. However, on the basis of human tolerances (see Figure 3) and assuming that human beings would not want to remain indoors constantly for long periods of time, it can be seen that *mean* daily temperatures of 104°F (40°C) and 14°F (−10°C) at the hottest and coldest seasons might represent reasonable limitations. It should be remembered that daily temperature extremes could typically go far above and below these mean daily values. Thus, these limits represent more rigorous conditions than might seem initially to be the case.

Furthermore, a reasonable fraction of a planet's surface area (say, at least 10 per cent) should possess the required conditions of temperature in order for it to be called habitable.

Regarding the temperature requirements, therefore, we specify that a region is habitable only if the mean annual temperature lies between 32°F (0°C) and 86°F (30°C), if the highest mean daily temperature during

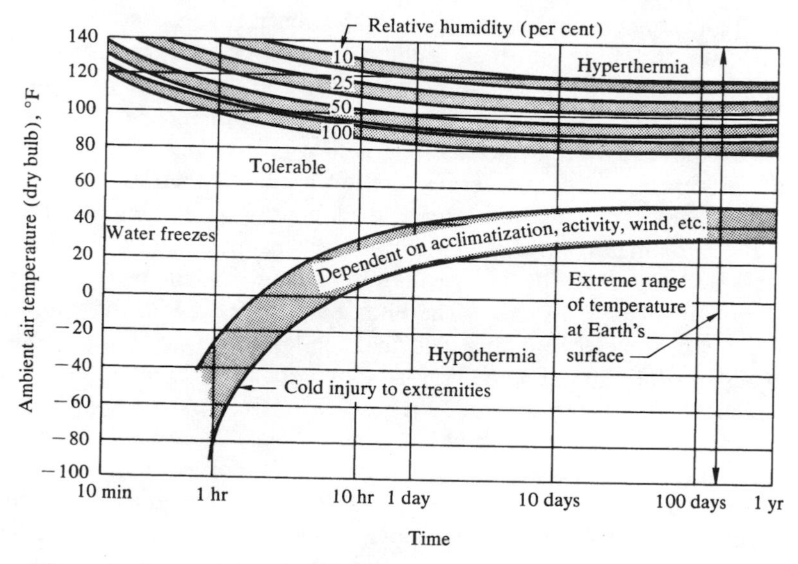

Figure 3. Approximate human time-temperature tolerances, assuming optimum clothing.

the warmest season is lower than 104°F (40°C), and if the lowest mean daily temperature of the coldest season is higher than 14°F (−10°C).

LIGHT

That portion of the electromagnetic spectrum visible to the human eye, which we call light, is contained principally between the wave lengths 380 and 760 millimicrons (mμ). By using very intense artificial sources, one can stretch the limits of human vision somewhat, obtaining the limits of 310 to 1050 millimicrons. Lying generally within this slightly wider range of wave lengths but mainly enclosed between 380 and 760 millimicrons, we find also the vision of other animals, the bending of plants toward light, the oriented movements of animals toward or away from light, and, most important, all types of photosynthesis. This is the domain of photobiology. According to Wald (1959), these same limits must be applicable everywhere in the universe.

Daily illumination intensities for active growth in green plants must fall between certain definite, but not too clearly established, limits. If the intensity of illumination is too low, for example, active photosynthesis can not proceed at a rate high enough to be useful; and if the intensity is too high, growth is inhibited by what has been termed "solarization." These lower and upper limits of illuminance may be set at approximately 0.02 and 30 lumens per square centimeter. (The maximum illuminance due to direct and scattered sunlight at the surface of the Earth is about 15 lumens per square centimeter.) The highest growth rates for terrestrial plants are encountered at intermediate levels of illumination. For some common species of algae, for example, the highest growth rates were found in the approximate range of 0.3 to 3.0 lumens per square centimeter (Krauss and Osretkar, 1961). On the other hand, human beings can see well enough to walk around if the illuminance is as low as 10^{-9} lumens per square centimeter, but they find that the level of over-all illuminance becomes painfully bright when it rises above about 50 lumens per square centimeter (Wulfeck *et al.*, 1958). This is an extreme oversimplification of a very complicated phenomenon, that of "glare discomfort," because many factors must be considered in addition to over-all illuminance: the reflectivity and distribution of surfaces and objects in the vicinity, the presence of shade and shadows, *et cetera.* Even ordinary levels of illuminance due to sunlight at the Earth's surface become intolerably high when one is surrounded by an unbroken field of diffusely reflecting material with a high albedo, such as fresh snow, giving rise to the well-known phenomenon of snow blindness.

[The above discussion of human tolerances for light refers to over-all illuminance levels, not to illumination directly entering the eye. Much lower values of illuminance must be specified in order for them to be considered tolerable to a man looking directly at a point source of light. For this, the upper limit is apparently of the order of 0.05 lumens per square centimeter (Hopkinson, 1956; Metcalf, 1958), which corresponds to a star of apparent visual magnitude about −21. The absolute lower limit of naked-eye detection of faint point sources in a very dark sky is about 10^{-13} lumens per square centimeter, which corresponds to a star of magnitude 8, although under the best typical viewing conditions, it is difficult to see stars fainter than magnitude 6.5.]

Thus, illumination requirements are set primarily by the needs of plants and are such that during the growing seasons, mean daytime illumination levels must lie between 0.02 and 30 lumens per square centimeter.

Another factor of great importance in affecting the growth of plants is the periodicity of illuminance. Especially in the temperate regions of the Earth, plant growth cycles are determined by the relative or absolute lengths of days and nights, as well as by temperature patterns. There is, of course, a close relationship between temperature and light on planets such as the Earth, which have normally transparent atmospheres and are illuminated by incandescent bodies such as the Sun. As will be brought out later, most habitable planets must receive their primary supplies of heat and light from the same type of source; hence, not too much can be made of the requirement for a certain level of light intensity as an independent variable.

GRAVITY

Experiments with human beings on large centrifuges have shown that relatively high levels of acceleration can be tolerated by some people for brief periods of time without permanent damage. For example, as indicated in Figure 4, accelerations of the order of 5 *g* (five times the

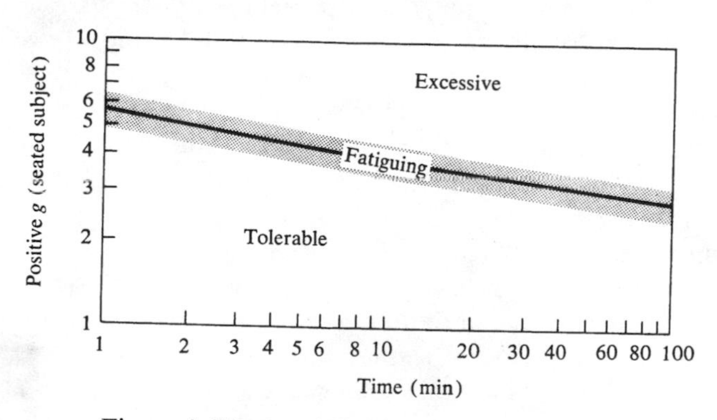

Figure 4. Human time tolerance: positive *g*.

normal gravity level at the surface of the Earth) can be tolerated by a seated man not wearing a "*g*-suit" for about 2 minutes without a blackout (loss of vision caused by an inadequate supply of blood at the eye level). An acceleration of 4 *g* can be tolerated for about 8 minutes, while 3 *g* have been tolerated for as long as an hour by some subjects in several experimental runs (Miller *et al.*, 1958). The subjects were seated and immobilized, however; they were not walking around or otherwise functioning in an everyday manner. At the conclusion of the 3-*g* experiments, the subjects reported quite pronounced muscular fatigue. Other experiments conducted in 1947 at the Mayo Clinic give at least an idea of the limitations imposed by increased gravitational fields (Code *et al.*,

1947). In these experiments, five human subjects were timed to see how rapidly they could scramble, creep, or crawl across the end of the centrifuge gondola, a distance of 7.5 feet, under various imposed accelerations. The results are shown in Table 1. The subjects were also timed to see how

Table 1. Time in Seconds Required To Move 7.5 Feet under Various Imposed Accelerations

g	1.00	1.41	2.24	3.16	4.12
Strongest subject	1.16	2.88	5.60	9.16	18.15
Weakest subject	1.85	7.11	14.85	21.83	...
Average of five subjects	1.51	4.88	9.36	15.80	...

quickly they could put on a standard parachute at various g levels. The average times required by three subjects were 17, 21, and 41 seconds for g levels of 1.00, 1.41, and 2.24, respectively.

When the averaged data are plotted on a graph (see Figure 5) on the basis of the time required to complete a given action relative to the time at 1 g, it may be seen that the relative time depends on the type of action. However, it may be concluded that the work required to perform various acts becomes excessive above approximately 2 g.

Animal experiments have pointed to similar conclusions. At the University of California, Smith and others have grown chickens in centrifuges for extended periods of time (Tobias and Slater, 1962). These chickens were able to survive prolonged exposure to accelerations up to 4 g, but lost weight unless the acceleration was less than 2.5 g. In these gravitational fields, the heart rate increased and the rate of respiration decreased. The life span of small animals also appears to decrease at gravitational forces higher than 2 g. Some mice showed increased life spans at 1.5 and 2.0 g (see Wunder *et al.*, 1962). Both plants and insects apparently can tolerate extremely high g levels (thousands of g). It must be acknowledged, however, that centrifuge experiments, with their necessarily high and physiologically disturbing angular velocities, do not faithfully reproduce the linear gravitational field of a massive planet.

On the basis of the available data, one might conclude that few people would choose to live on a planet where the surface gravity was greater than 1.25 or 1.50 g. It is true that many people who are 25 to 50 per cent overweight live very normal lives and manage to accomplish as much as, or more than, many people whose weights conform more closely to the standards for their heights and ages. On the other hand, it is also generally true that physical activity is more exhausting to people who are carrying excessive burdens of fat. Overweight, of course, can not be equated with increased *g* on a one-to-one basis, but the relationship may give us a notion of the performance to be expected from human beings living on planets with gravitational accelerations greater than that of the Earth.

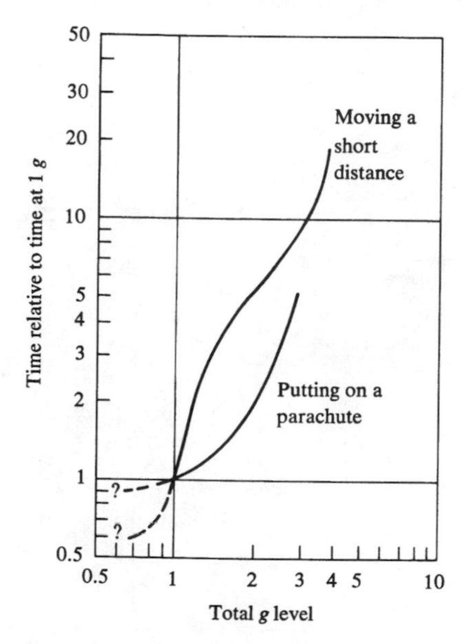

Figure 5. Time required to complete a given act relative to the time required at 1 *g*.

There does not seem to be a corresponding lower gravitational limit on the tolerances of human beings; that is, there is no conclusive evidence that a certain level of gravity *per se* is required for their normal physiological functioning. From Figure 5 it may be inferred that activities in reduced-*g* fields would take somewhat less time (and energy) than in fields of 1 *g*. More data on this can be expected in the near future from manned-space-flight programs now projected.

ATMOSPHERIC COMPOSITION AND PRESSURE

A habitable planet must, of course, have a breathable atmosphere; and this now can be rather completely specified in terms of its component gases and their concentrations or partial pressures (see Figure 6).

As far as we know, the only essential ingredients of a breathable atmosphere are oxygen and minor amounts of water vapor. The inspired partial pressure of oxygen must lie between two extreme limits: a lower limit below which hypoxia is encountered and an upper limit above which oxygen toxicity becomes a problem.

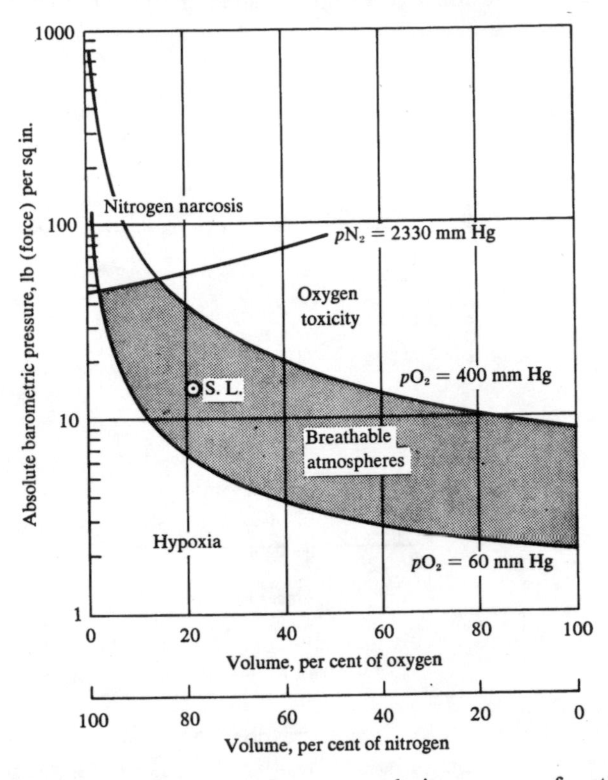

Figure 6. Breathable mixtures of oxygen and nitrogen as a function of barometric pressure.

In determining the level of inspired partial pressure of oxygen (or any gas) under any specified condition of composition and total barometric pressure, it is necessary to correct for the fact that as air is inhaled it is also humidified in the nasal passages and throat, so that, by the time it reaches the lungs, it is normally saturated with water vapor at body temperature. For this reason, a correction is made:

$$pO_2 \text{ (inspired)} = (pB - pH_2O) \times F_{O_2},$$

in which pO_2 is the partial pressure of oxygen in the air entering the lungs, pB is the total barometric (ambient) pressure, pH_2O is the partial pressure of water vapor in the air entering the lungs, F_{O_2} is the fraction (by volume) of oxygen in the ambient air. For pressures expressed in millimeters of mercury (mm of Hg), pH_2O is assumed to be 47 millimeters of mercury. Thus, under normal sea-level conditions where barometric pressure is 760 millimeters of mercury, $pO_2 = (760 - 47) \times 0.2093 = 149$ millimeters of mercury.

The lower limit of inspired oxygen partial pressure is approached by the inhabitants of a mining settlement at Aucanquilcha in the Chilean Andes, situated at an altitude of 17,500 feet above sea level. This is said to be the greatest altitude at which men are known to live permanently (Pugh and Ward, 1954). It is considerably higher than the environment of the Tibetans, most of whom reside and work their land at altitudes between 12,000 and 16,000 feet. At 17,500 feet the inspired partial pressure of oxygen is about 72 millimeters of mercury, yet the miners of Aucanquilcha lead very strenuous lives and appear to be completely acclimated to the low level of oxygen pressure. To reach the entrance of the mines where they work, they climb 1500 feet each day to an altitude of 19,000 feet, at which the inspired partial pressure of oxygen is 68 millimeters of mercury. Yet even these conditions probably do not represent the ultimate lower level of oxygen pressure that can be tolerated as a steady-state condition by some men. According to mountain climber N. E. Odell, ". . . our evidence has shown us emphatically that one can live and feel fit for an indefinite period at 23,000 feet [pO_2 (inspired) = 53 millimeters of mercury] . . .." (See Norton, 1925.)

The upper limit of inspired oxygen partial pressure has been found experimentally to be approximately 400 millimeters of mercury (equivalent to about 56 per cent oxygen in the air at sea-level pressure). This limit is approached in the therapeutic use of oxygen in hospitals, where the accepted ceiling is 40 per cent oxygen, to be on the safe side (Ingalls, 1955). It was discovered in the 1950's that blindness in premature babies (retrolental fibroplasia) often resulted from the use of excessively high oxygen concentrations in their incubators. Since human beings vary a great deal in their tolerances of environmental extremes and even in their ability to adapt, it is not be to be expected that a single set of limits would apply to all people. Rather, it is desired here to set limits such that at least some people could live within them. Following this philosophy, for our purposes we may state that the inspired partial pressure of oxygen must be greater than about 60 millimeters of mercury but less than about 400 millimeters of mercury.

An inspired partial pressure of oxygen of 60 millimeters of mercury with an atmosphere of pure oxygen corresponds to a total barometric pressure of 107 millimeters of mercury, equivalent to 2.07 pounds per square inch absolute (psia). At barometric pressures slightly below this level, gaseous swelling of the body due to the formation of bubbles in the blood has been observed in experiments with animal and human subjects. Carbon dioxide and water vapor are believed to be the main gases involved in the swelling phenomenon (Wilson, 1961).

There are only certain diluents that may be mixed with the oxygen in a breathable atmosphere and each has an upper limit of inspired partial pressure that should not be exceeded. Symptoms of narcosis due to nitrogen, argon, krypton, and xenon (all chemically inert gases) have been reported when inspired partial pressures exceed certain levels. Xenon, in fact, has even been used as an anesthetic in surgical operations: an 80-per-cent xenon and 20-per-cent oxygen mixture, at 1 atmosphere pressure, will produce unconsciousness in 3 to 5 minutes (Cullen, 1951). The narcotic effects of carbon dioxide are even more widely known. Presumably neon and helium, and possibly hydrogen, also have narcotic effects at sufficiently elevated pressure levels. Table 2 shows the approximate upper limits for each of the permitted inert diluents in a breathable atmosphere.

Table 2. Approximate Upper Limits for Permitted Inert Diluents in a Breathable Atmosphere

Diluent	Approximate maximum permissible inspired partial pressure (mm of Hg)
Hydrogen	...
Helium	61,000 (?)
Neon	3,900 (?)
Nitrogen	2,330
Argon	1,220
Krypton	350
Xenon	160
Carbon dioxide	7

Hydrogen is a special case in that only noncombustible mixtures of hydrogen and oxygen could be regarded as acceptable and one would never expect to find *both* free hydrogen and free oxygen simultaneously present in a planetary atmosphere. The figures for helium and neon are extrapolated. None of the gases listed is known to be necessary to make up a breathable atmosphere; hence, presumably, any or all can be completely absent as long as the oxygen partial pressure falls within the proper range. Really prolonged tests on people living in atmospheres containing no inert gases have not yet been carried out, however, so it can not be stated categorically that inert gases are unnecessary. (The longest test carried out on human beings in atmospheres containing no inert gases has been of 17 days' duration.) Human evolution has taken place in an atmosphere containing almost 80 per cent inert ingredients, and it may be that some inert gas fraction is needed for the proper functioning of the respiratory system during critical periods of life. Atelectasis (the collapse of mucous-clogged lung alveoli in certain respiratory illnesses, due to the absorption of the contained gases), for example, would be more apt to occur if there were no inert gases in the air being breathed. Also, since carbon dioxide is needed by plants, a lower limit is required on the partial pressure of this constituent. The normal concentration of carbon dioxide in the Earth's atmosphere is 0.03 per cent, equivalent to a partial pressure of 0.21 millimeters of mercury. A reasonable minimal value for supporting normal plant life has not been determined, but possibly it would be of the order of 0.05 to 0.10 millimeters of mercury. Some nitrogen is also needed to supply nitrogen compounds to plants and animals. The minimum necessary amount is probably small but is not known. A small fraction of the free nitrogen in the Earth's atmosphere is constantly being converted into the oxides of nitrogen by lightning flashes (over 3 billion lightning strokes per year, according to the Lightning Protection Institute of Chicago) producing some 100 million tons of fixed nitrogen annually; without this, plant life on the planet would not be able to obtain adequate supplies of available nitrogen. Nitrogen fixation is also accomplished by bacteria attached to the roots of certain leguminous plants well known to farmers all over the world.

Only traces of other gases could be tolerated in the atmosphere of a habitable planet. Upper limits on the tolerable concentrations of some naturally occurring or elemental gases are shown in Table 3.

All of the gases listed in Table 3 are quite active chemically. Thus, on a planet with an atmosphere containing both free oxygen and water vapor, one would not expect to find any of them present as permanent constituents, except in trace amounts. Methane in low concentration is not generally regarded as a toxic material, but it would oxidize slowly and tend to disappear from an atmosphere containing oxygen.

There are many other toxic gases with which we need not be concerned, since there is no reason to expect them to occur naturally in any important concentration on the surface of any planet. Tables 2 and 3 include all of the common gases that might potentially be found in planetary

Table 3. Tolerable Concentrations of Selected Gases[a]

Gas	Threshold limits (parts per million by volume at 1 atmosphere pressure)
Ammonia, NH_3	100
Carbon monoxide, CO	100
Chlorine, Cl_2	1
Fluorine, F_2	0.1
Formaldehyde, HCHO	10
Hydrogen chloride, HCl	5
Hydrogen cyanide, HCN	10
Hydrogen sulfide, H_2S	20
Methane, CH_4	50,000[b]
Nitrogen dioxide, $2NO_2 \rightleftharpoons N_2O_4$	25
Nitrous oxide, N_2O	. . .[c]
Ozone, O_3	0.1
Sulfur dioxide, SO_2	5

[a] All gases listed are known to occur free in nature at times, except chlorine, fluorine, formaldehyde, and hydrogen cyanide.

[b] Lower inflammability limit in sea-level air; toxic limit is at a higher concentration.

[c] 20 to 40 per cent in air induces analgesia in less than 5 minutes.

atmospheres, with the exception of water vapor. Water is a special case in many ways. For one thing, it is the only common material with a freezing point in the acceptable range of temperature and pressure. Because of its special characteristics it requires treatment in a separate section.

From Figure 6 it may be seen that there is a lower limit to the atmospheric pressure required on a habitable planet: a barometric pressure of about 2.1 pounds per square inch for an atmosphere of pure oxygen. At the other extreme, the maximum tolerable total barometric pressure for human beings has not yet been determined. An atmosphere containing 2 per cent oxygen and 98 per cent helium at a total pressure of 150 pounds per square inch, for example, falls within the tolerable region, although really prolonged exposures have not been studied experimentally. Conceptually, however, there is a maximal total pressure beyond human tolerance that would be reached when the gas density had increased to a point such that there would be highly turbulent flow in the air passages of the nose, and the work of breathing would become exhausting. Donald is reported as stating that at a pressure of 8 atmospheres (118 pounds per square inch absolute) the turbulence is so great that one can actually feel eddy currents in the air as it flows through the mouth (Otis and Bembower, 1949).

To summarize, then, the atmosphere of a habitable planet must contain oxygen with an inspired partial pressure between 60 and 400 millimeters of mercury and carbon dioxide with a partial pressure roughly between 0.05 and 7 millimeters of mercury. In addition, the partial pressures of the inert gases must be below certain specified limits and other toxic gases must not be present in more than trace amounts. Some nitrogen must be present so that nitrogen in combined form can find its way into plants.

WATER

Water is, beyond doubt, one of the most remarkable substances in the universe and the one most inextricably linked with life of all kinds. A classic essay on the many peculiar properties of water that make it uniquely suited to the requirements of living things is included in *The Fitness of the Environment* by L. J. Henderson (1958). Among other things, Henderson discusses the remarkable heat-regulating properties of water (its extraordinarily high heat of vaporization and its high heat of fusion), its anomalous expansion when cooled below 4°C, the low density of ice, the incomparable powers of water as a solvent, its high dielectric constant, and its high surface tension. In a lighter vein, even James Joyce in *Ulysses* expounded on the marvelous properties of water.

Man and his entire ecology are utterly dependent on water. It can be said categorically that a habitable planet must have fairly large open bodies of liquid water, for without oceans there could be no extensive precipitation and hence no salt-free ground water to provide supplies of fresh water. Precisely what ratio of ocean surface area to total planetary surface area is necessary may be very difficult to determine. It is clear that a certain critical total quantity of water is necessary on the surface of a planet before bodies of water can appear. If there were less than this amount, then all of the water would be in the form of water vapor or watei absorbed on the surfaces or held interstitially between the solid particles of rock of which the crust is composed. On the other hand, a planet completely covered with water, a pelagic planet without permanent dry land, could hardly be considered habitable from man's point of view. On the Earth, a multiplication of our present water supply by a factor of four would be required for complete inundation of all our continents.

Humidity in a breathable atmosphere is also highly important. The uncomfortable effects of high levels of humidity at high temperatures are well known; but there may also be adverse physiological effects due to extremely low levels of water vapor pressure in the air, particularly at the higher temperatures. This latter condition causes very rapid drying of the mucous membranes of the nose, mouth, and throat; and continuous exposure to very low levels of water vapor pressure might well be intolerable.

Regarding water, then, a habitable planet must be able to retain it and must have open bodies of fluid water on its surface, but the proportion of surface area covered by water must be somewhat less than 90 per cent.

OTHER REQUIREMENTS

The above-stated requirements for conditions of temperature, light, gravity, atmospheric composition and pressure, and water are probably the major human requisites; yet there are many others. These will be stated briefly.

As indicated earlier, *other life forms* must be present, since ultimately all human food supplies (and probably the very existence of free oxygen in the atmosphere) depend on photosynthesis in green plants. There must also be an absence of inimical intelligent beings in prior possession, for the phrase "habitable planets for man" is meant to indicate planets that are not already "taken." Man, presumably, can always cope with nonintelligent life forms.

Commonly experienced *wind velocities* in otherwise habitable regions must be of tolerable levels. Regions in which wind velocities consistently reach strong gale force (about 50 miles per hour) or higher would not be considered habitable.

Similarly, *dust* normally encountered should be below certain specified levels. Air Force Pamphlet 160-6-1 suggests that total dust (containing less than 5 per cent free silica) should not exceed 50 million particles per cubic foot of air and that high-silica dust (containing more than 50 per cent free silica) should not exceed 5 million particles per cubic foot of air. Dust concentrations exceeding these threshold limits are considered harmful.

The bodies of water of a planet are the primary receptacles for airborne dust; and water droplets forming on dust nuclei provide the primary means of removing dust from the atmosphere. Thus, it is to be expected that a planet possessing oceans would not have a particularly dusty atmosphere, while a planet with a turbulent atmosphere but without oceans would be a very dusty place indeed.

Ambient levels of *radioactivity* or ionizing radiation, whether caused by radioactive materials in the crust or by high-energy particles coming through the atmosphere from stellar sources, must be of acceptable intensity. For genetic reasons it would probably be desirable to specify dosages from natural background radiation of less than 1 roentgen per year on a steady-state basis, or approximately 0.02 rem (roentgens-equivalent, man) per week. (The Atomic Energy Commission's steady-state tolerance level is 0.3 rem per week for workers in atomic energy plants; the average natural background radiation on the Earth's surface is about 0.003 rem per week.) Not much is known at present about the effects of prolonged steady-state exposures to very low levels of dose rate. The "mutation-doubling rate," however, is generally taken as 30 to 50 roentgens; and if

Table 4. Summary of Characteristics of Habitable Portions of Planets

Characteristic	Tolerable range		
Temperature			
mean annual (°F)	32	to	86
mean daily (°F)	14	to	104
Light at surface, peak values (lu/cm^2)	0.02	to	30
Gravity (relative to Earth normal)	...	to	1.5
Atmospheric composition, inspired partial pressure (mm of Hg)			
oxygen	60	to	400
carbon dioxide	0.05	to	7
helium[a]	0	to	61,000 (?)
neon[a]	0	to	3,900 (?)
nitrogen[a]	10 (?)	to	2,330
argon[a]	0	to	1,220
krypton[a]	0	to	350
xenon[a]	0	to	160
toxic gases	trace amounts only[b]		
Water vapor, partial pressure (mm of Hg)	...	to	25[c]
Other characteristics			
open bodies of water			
indigenous life forms			
tolerable wind velocities, dust levels, natural radioactivity, meteorite-infall rates, volcanic activity, and electrical activity			

[a] If more than one inert ingredient is present, maximal values should be in proportion to the relative concentrations.

[b] Amount depends on specific gas.

[c] A function of temperature.

this amount is spread out over a human lifetime up to the end of the reproductive period, such a dose would be accumulated at about 1 rem per year. Much higher dose rates are tolerable, of course, and probably the present mutation rate in human beings could be multiplied by a factor much higher than two before it would become painfully evident.

Other conditions that might render a planet uninhabitable would be an excessively high *meteorite-infall rate*, an excessive degree of *volcanism*, a high frequency of *earthquakes*, and possibly an excessive degree of *electrical activity* (lightning).

In summary, habitable portions of planets have the characteristics shown in Table 4.

Having now summarized the principal requirements of human beings with respect to the environmental conditions provided by a planet, we must next consider the varieties of planets that are known to exist or that can be postulated. Once the ranges of environments that prevail on planetary surfaces are circumscribed, we can determine how well these correspond to human requirements.

CHAPTER 3

Introduction to General Planetology

About 50 years ago, it was recognized by E. Hertzsprung (1911) and H. N. Russell (1914) that the stars may be grouped into families by spectral type (temperature) and luminosity (absolute magnitude). Since the time of their discovery of some of the fundamental stellar properties, many general relationships between mass, luminosity, age, diameter (density), temperature, spectral type, composition, internal conditions, nuclear reactions, *et cetera*, have been deduced for stars. Ideas are continuing to be developed about the evolution of stars and the internal and external physical and chemical changes accompanying their aging processes. Concepts are being formed about the modes of formation of stars, the relationships between members of close binaries, the distribution of star types in galactic clusters, *et cetera*. It is not necessarily true that all these ideas are correct; many, in fact, are conflicting. The point is that the large, luminous bodies of matter in the universe called stars or suns are not individually unique and curious specimens, seemingly completely unrelated to one another; rather, they are now recognized to be members of a class with familial similarities. Although differing in mass and age, when their intrinsic differences are taken into account, the class of stars appears to form a continuum wherein all their other observable properties seem to be related to the primary accidents of mass, age, rate of rotation, propinquity to other massive bodies, and, possibly, primordial chemical composition.

The number of luminous bodies individually detectable in the sky runs into the millions, of which some 500,000 may be called well-observed stars. Consequently we have available a large population to study and compare, to analyze statistically for number and distribution, to analyze spectroscopically, and so on.

Quite a different situation prevails when it comes to the bodies in the universe that are not self-luminous. For this class of bodies, the existence of which may be known only because of their ability to reflect light or be inferred from observed oscillations of nearby self-luminous bodies, the known members are primarily a few relatively small bodies within our solar system, that is, the planets, satellites, and asteroids.

Because the planets of the solar system are so few in number, they have usually been treated as unique objects and studied as such. Yet, conceptually, if some current ideas about the formation of stars are substantially correct, the number of stars in the universe may be exceeded by the number of bodies that are not self-luminous. It is to be expected that the class of bodies that are not self-luminous could be subdivided into families or classified in a number of ways according to physical or positional properties. From this point of view, then, the planets of the solar system might be regarded, not as unique specimens, but as members of a large family of such objects, the totality of which would form a continuum in which the relationships between the definable physical characteristics would be found to follow certain general laws of nature.

The word "planetology" has been used in the past to mean the "study and interpretation of surface markings of planets and satellites," where a planet is defined as "any body, except a comet or a meteor, that revolves about the sun of our solar system."* A more general term is needed to cover the physical properties of all non-self-luminous bodies, whether they are a part of our own solar system or are orbiting about some other star. To embrace this usage, the term "general planetology" is defined here as "a branch of astronomy that deals with the study and interpretation of the physical and chemical properties of planets." In this context, planets will then be defined as "massive aggregates of matter that are not large enough to sustain thermonuclear reactions in their interiors."

General planetology, then, is concerned primarily with deducing the interrelationships among the various properties of planet-like objects, roughly those objects with a mass less than about 10^{31} to 10^{32} grams (that is, less than about $\frac{1}{200}$ to $\frac{1}{20}$ the mass of the Sun). A number of intrinsic, positional, and resultant (or secondary) properties of planetary objects are itemized below.

Intrinsic properties include mass, rate of rotation, and age.

Positional properties include mean distance from primary star (the star about which the planet orbits), inclination of equator to orbital plane, orbital eccentricity, relationships with other planetary bodies (satellites,

---

* *Webster's New International Dictionary* (2nd ed., unabridged). Springfield: G. & C. Merriam Co., 1951.

et cetera), and properties of the primary star (or star system) in which the planet exists.

Resultant properties include intensity and type of radiation received from primary (or primaries); surface temperature patterns; gravitational force at solid surface; atmospheric composition, density gradient, and temperature profile; internal structure and composition; atmospheric pressure at solid or liquid surface; atmospheric circulation patterns; tidal factors; radioactivity levels; existence of life forms; volcanic activity; and meteorite-infall rate.

If we look on the planetary bodies in the solar system now as members of a family and search for the familial resemblances, a number of interesting patterns emerge. Some of these will be discussed later in detail. Worthy of mention here are the relationships between mean density and mass; between rotational energy per unit mass and planetary mass; and among the possible atmospheric composition, the mass, and the intensity of the radiation field. Some other interesting relationships may be inferred from empirical extrapolations, or interpolations, or as perturbations from conditions known to exist on the Earth. Still others may be calculated from theoretical considerations.

At present it is obvious that our knowledge of the underlying laws of general planetology must be far from complete. For one thing, many of the data on the planets of the solar system are known only approximately (although they are often reported to three significant figures); hence certain data are not reliable enough to use in making correlations. The density of Mercury, for example, has been reported in the literature as being as low as 3.7 grams per cubic centimeter and as high as 6.2 grams per cubic centimeter (Firsoff, 1952); numerous intermediate values have also been given. This variation in reported density is a consequence of the extreme difficulty of measuring with precision either the mass or the diameter of Mercury. The physical dimensions and densities of Uranus and Neptune are also only very roughly known. In addition, our current knowledge concerning the properties and behavior of ordinary matter under extreme conditions of pressure (as must exist in the interiors of planets) is still quite rudimentary. It can not yet be said that we have anything but a fragmentary picture of the causes of mountain-building processes, earthquakes, and volcanoes, or of the structure of the Earth's crust, mantle, and interior.

Finally, there are many areas of study in which the complete elucidation of all the effects that could take place is so extraordinarily difficult and complex with the use of present techniques that it is often necessary to simplify the problems greatly in order to attack them at all. A case in point is the current state of Earth meteorology. Great strides are being made in the understanding of winds, storms, precipitation, air-circulation patterns, and other atmospheric phenomena; but many questions still remain unsolved (for example, the cause of the ice ages) and most weather prediction must still be done on a chiefly empirical basis. Yet, how much more difficult it would be to elucidate planetary meteorology in a completely general manner, taking into account every possible combination of atmospheric composition, mass, surface gravity, rate of rotation, land-sea ratio, tilt of axis, *et cetera.*

There are certain overriding or dominant astronomical factors, however, that permit estimations of the grosser aspects of planetary meteorology to be made. For example, the conditions necessary for the loss or retention of various gaseous atmospheric constituents can be estimated, and thus certain subclassifications of planets may be designated approximately. A knowledge of the relative universal abundances of the chemical elements, coupled with a knowledge of the chemical and physical properties of the most abundant elements and compounds, permits deductions concerning the atmospheric constituents that need to be considered.

The main objective of general planetology, in common with all science, is a fuller understanding of the universe in which we live. Some subsidiary objectives are to define the general characteristics of planetary systems; to obtain estimates of the number of planetary systems; to gain a clearer understanding of the characteristics of habitable planets and to obtain a more definitive estimate of the number of habitable planets in our Galaxy and in the universe; to indicate which of the stars in the neighborhood of the Sun would be most likely to possess habitable planets and what the probabilities might be; and finally, to obtain a better understanding of our own planet and an appreciation of the combination of factors that makes the Earth a comfortable place to live.

The following sections will deal specifically with the intrinsic and positional parameters of planets and some of the relationships that have been found or can be inferred. None of this material is absolute, above criticism, or secure from future modification or revision; but many interesting conclusions may be drawn, given even the present incomplete development of general planetology.

SOME GENERAL PROPERTIES OF THE SOLAR-SYSTEM PLANETS

Properly, any complete discussion of the properties of massive aggregations of matter in the universe should commence with a description of the modes of formation of stars and planets. This whole subject, however, is a controversial one--and somewhat outside the scope of the present chapter. Harlow Shapley (1959) lists fifteen different ancient and modern hypotheses concerning planetary origin, indicating a preference for the fifteenth, his "hypothesis of desperation," or the Survival of the Conforming (that order comes out of chaos simply because the less stable members are swept up or otherwise removed by the more stable members so that, after a long time, only the most stable members remain). It is similar in some respects to the theories of dust and gas accretion put forth in recent years by von Weizsäcker (1944), Urey (1952), Kuiper (1951), Hoyle (1950, 1955), and others. According to such theories, stars originate by accretion of dust and gas within clouds of cosmic matter, and planets are formed from the leftover matter surrounding newly formed stars. Much has been written about the details of formation of stars and planetary systems. The "accretion" hypothesis seems to account quite satisfactorily for most of the observed dynamic and physical properties of the bodies of the solar system (as no other hypotheses do). Thus, without going into the minute details, about which there is much controversy, we will assume here that the "accretion" hypothesis is correct. Our present objective is to point out some of the intrinsic properties of massive aggregations of matter.

Classification of Celestial Bodies on the Basis of Mass. By far the most important intrinsic property of large aggregations of matter is *mass*. If we define stars as aggregations of matter massive enough to sustain thermonuclear reactions in their interiors now or at some time in their past, we may define planets as aggregations of matter that are not now, and never were, massive enough to sustain such thermonuclear reactions.

All of the stars of which we have any observational knowledge are in the mass range from 0.04* to about 60 times the mass of our Sun, although the vast majority of the stars that we can observe have 0.2 to 5 times the mass of our Sun. Eddington defined a star as any aggregation of matter with a mass between 10^{32} and 10^{37} grams (0.05 to 5000 solar masses), since he believed that a body less than 10^{32} grams in mass could not remain self-luminous as a star, and material aggregations exceeding 10^{37} grams in mass could not survive the instability due to radiation pressure that would tend to blow them to pieces.

Clearly, there is some upper limit to the mass of a planet and a lower limit to the mass of a star, although the exact value of mass at which this transition takes place in not known. The existence of bodies having masses in the transition region is not subject to direct confirmation at present. Present indications are, however, that the transition region lies somewhere

* This is the mass determined for each of two stars designated L726–8A and L726–8B.

between 10^3 and 10^4 Earth masses (6×10^{30} to 6×10^{31} grams, or 3×10^{-3} to 3×10^{-2} solar mass). Figure 7 shows the mass-volume relationship both for self-luminous and non-self-luminous bodies over the entire range of known celestial bodies (except for diffuse aggregations such as galactic dust, gas clouds, and comets).

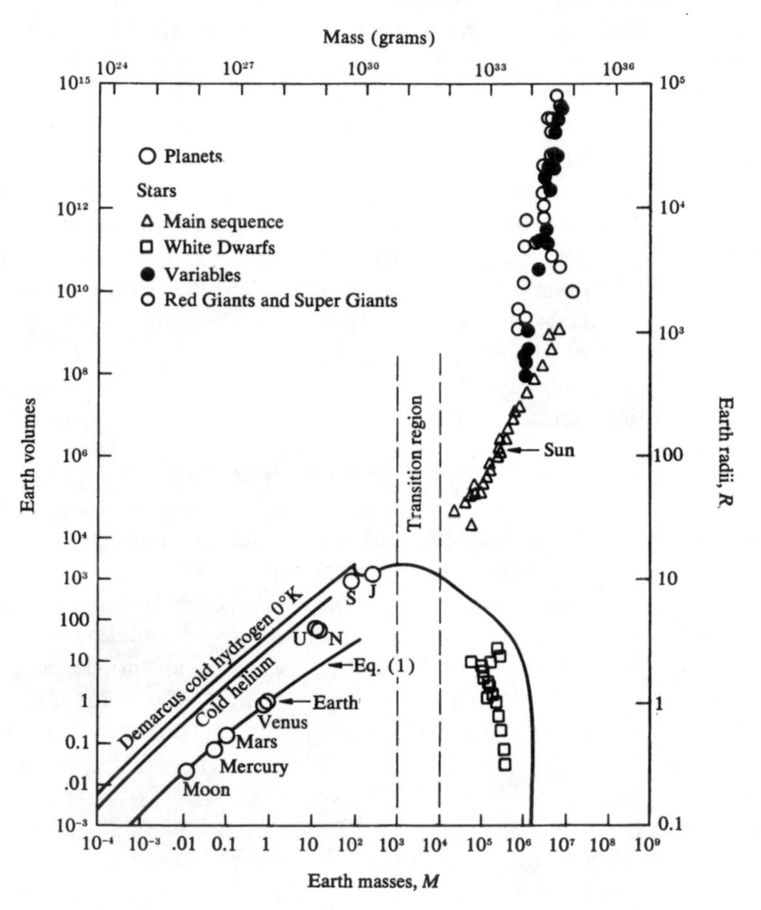

Figure 7. Mass-volume relationships of stars and planets.

The dividing line between the largest planet and the smallest star may not be very sharp; some large planets may even be slightly more massive than some small stars. It may depend to some extent on the density of matter available at the time of formation and the relative velocity of matter in the vicinity. If the rate of the matter accumulation by a nucleus depends on its instantaneous mass, the local density of matter, and the relative velocity of matter, then it follows that two objects having the same final mass may have had different rates of growth. A very slow rate of growth, permitting a longer time to radiate away the kinetic energy resulting from the infall of its constituent particles, might result in central temperatures not quite high enough to trigger the thermonuclear reactions necessary for a star. A higher rate of growth, on the other hand, would result in the formation of a star. A borderline class of objects might exist: objects that are just able to trigger thermonuclear reactions, but lose this ability on expansion and thus oscillate or pulsate weakly on the border between planet and star.

The growth rate of a planet would be strongly affected by the presence of larger objects in the same system. These, by rapidly adding to their own mass, would be gradually reducing the mean density of the diffuse matter available to the planet. The farther away the orbit of a planet is from the central body, all other things being equal, the less the two tend to compete for growth material, and the lower is the mean velocity of local material passing near the planet. Thus the planet's effective capture radius, and hence its growth rate, is greater than that of an object of the same instantaneous mass orbiting nearer the central star.

Figure 7 suggests an approximate upper mass limit for planets. The lower mass limit for planets must be somewhat arbitrary. Or, to put it another way, as one looks at bodies of smaller and smaller mass, at what point does one begin to describe them as asteroids, meteoroids, or pieces of rock? Properties of masses in the lower range of mass are discussed below.

If one displays graphically the relationship between dimensions and density for all the terrestrial bodies in the solar system for which reliable data are available, an interesting pattern emerges. This pattern, shown in Figure 8, indicates that the mean densities of the Earth, of Mars, of the Moon, and of average material on the Earth's surface (uncompressed rock) all fall on a curve of the form

$$\bar{\rho} = \rho_0 \, e^{AR}, \tag{1}$$

where $\bar{\rho}$ = mean density, grams per cubic centimeter;

ρ_0 = surface density = 2.770 grams per cubic centimeter;

A = a dimensionless constant = 0.6904;

R = mean radius of rock sphere, relative to mean Earth radius.

That the observational points for Venus fall below the curve is easily understood, since the observed radius of Venus includes the depth of atmosphere beneath the opaque cloud layer. Hence the rocky sphere beneath the cloud must have a smaller radius and a higher density than is indicated by the observational points.

The points for Mercury are quite scattered because of the great difficulty in measuring precisely either its mass or dimensions. Whether the relatively high densities that have been determined by some observers represent an anomaly associated with Mercury's proximity to the Sun, or are due to some systematic errors, is not known at this time.

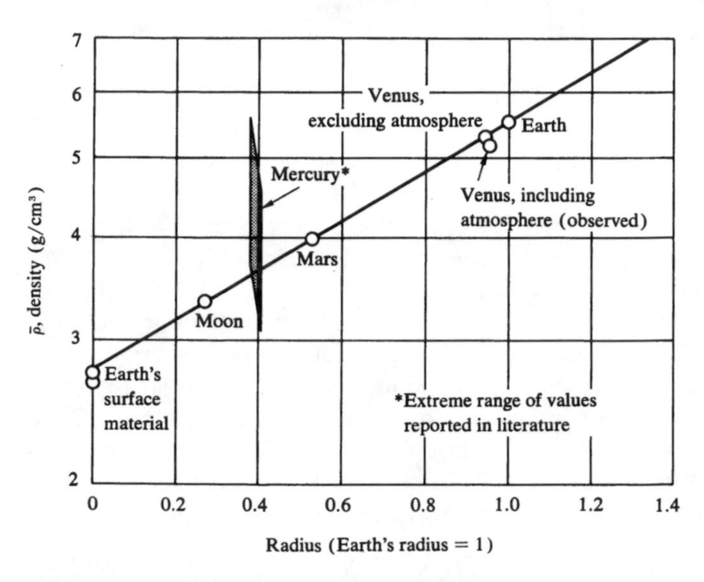

Figure 8. Density-radius relationship of terrestrial planets.

It is possible that the high density of Mercury (if it is indeed exceptionally high) could be attributed to extreme degassing and loss of water of crystallization from its rocky substance to a great depth below the surface. Since Mercury apparently keeps one face perpetually toward the Sun, the surface-rock temperature on the sunny face may reach about 700°C (Newburn, 1961). Temperatures of this order of magnitude, accompanied by high vacuum, can effect loss of water of crystallization from some common rock minerals and thus produce increased density. Another hypothesis is that the rock materials of which Mercury is composed were degassed before their aggregation into a planet.

Data on other terrestrial-type bodies in the solar system, for example, the large satellites of Jupiter and Saturn, are not sufficiently reliable to warrant their inclusion in Figure 8.

If the above empirical relationship is accepted as a working hypothesis *for bodies without extensive atmospheres,* then some interesting subsidiary properties may be calculated. Figure 9 shows values for surface gravity at the solid surface, for velocity of escape, and for mass, each plotted against radius. From some deductions about the structural strength of uncompressed materials and, with Kopal's data on the mountains of the Moon (Kopal and Fielder, 1959), these bodies may be further classified as approximately spherical aggregations or as bodies of highly irregular

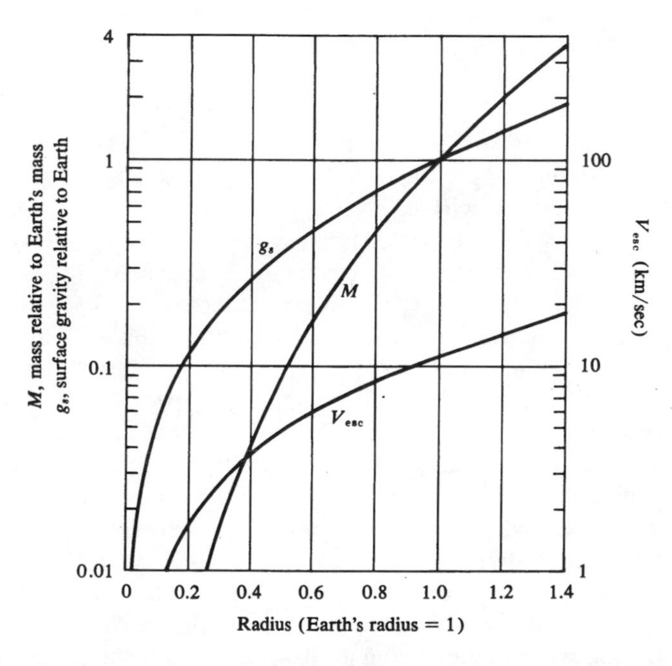

Figure 9. Characteristics of terrestrial planets: mass, gravity, and escape velocity versus radius.

shape. Small chunks of rock can exist in almost any conceivable shape; but a large accumulation of matter tends to become spherical because the forces of its own gravity deform it, making loose particles roll downhill, as it were, and making even the strongest rigid materials flow like liquids until an equilibrium has been established. An example of this may be seen in the icecap of Greenland, where snow keeps accumulating and compacting into ice at the center of the island, yet the ice does not keep rising. This is because it flows under its own weight and eventually breaks off into icebergs as it flows to the edge. Similarly, it would be impossible on the Earth's surface to build a structure that exceeded a certain height, no matter what material was used. Once this height had been attained, the addition of more material would cause the structure to sink into its own base or flow out at the edges. The equilibrium height could not be exceeded.

The greatest possible mass of a body capable of preserving a highly irregular shape is about 10^{-5} Earth mass (about 10^{23} grams). For materials with higher yield strengths, the transition from irregular to more nearly spherical shapes may take place at slightly higher values of mass. Unrealistically high yield strengths must be assumed, however, to preserve grossly nonspherical shapes for masses as large as 10^{-4} Earth mass (about 10^{24} grams).

Some anthropocentric characteristics of rock masses are as follows: A man can jump off any body having a mass less than about 7×10^{17} grams (mean radius, approximately 2.4 miles) if we assume that he can jump with an initial velocity of 16 feet per second. A man can throw an object such as a baseball completely away from any body having a mass less than about 2×10^{20} grams (mean radius, approximately 16 miles), assuming that he can launch the object at about 110 feet per second. A rifle bullet can be shot away from any body having a mass less than about 3×10^{24} grams (radius, approximately 400 miles), assuming a muzzle velocity of 2700 feet per second. In fact, people exploring small asteroidal bodies in this size range will have to be careful about throwing or shooting, for objects launched horizontally at velocities greater than circular orbital velocity but less than escape velocity will remain in orbit indefinitely and could consititute a hazard to personnel each time they skimmed back in to make their closest approach (periplanet).

Because small massive bodies have so little ability to capture small particles passing near them, it is to be expected that their rates of growth by aggregation would be exceedingly low, unless the density of material in their vicinity were high. From consideration of theoretical growth rates of bodies imbedded in a tenuous gas and dust cloud, it may be possible to conclude something about final size and the time required to reach this final size. Large bodies would tend to grow more rapidly as their mass increased, thus depleting the reservoir of material available for the growth of bodies that were initiated at a later time.

It is interesting that the mass-number relationship for bodies in the solar system shows a break at a mass of about 10^{-6} Earth mass (see Figure 10). There are a great many bodies having a mass less than 10^{-6} Earth mass (the number apparently increasing exponentially with decreasing mass, based on asteroid and meteorite observations); but there are only a few bodies of mass greater than 10^{-5} Earth mass.

Using the above information, it is possible to set an approximate lower mass limit for planets at about 10^{-5} Earth mass. Massive bodies smaller than this might be termed asteroids or meteoroids or rocks or stones, but not planets. It will be noted that the word, planet, as used here includes objects that are either planets or satellites in the solar system. As we are not dealing at the moment with positional considerations but only with the intrinsic properties of planets, it will be understood that the word, planet, signifies any massive body between the mass limits of roughly 10^{-5} to 10^4 Earth masses, whether it is isolated in space or is orbiting about a star or about another planetary body.

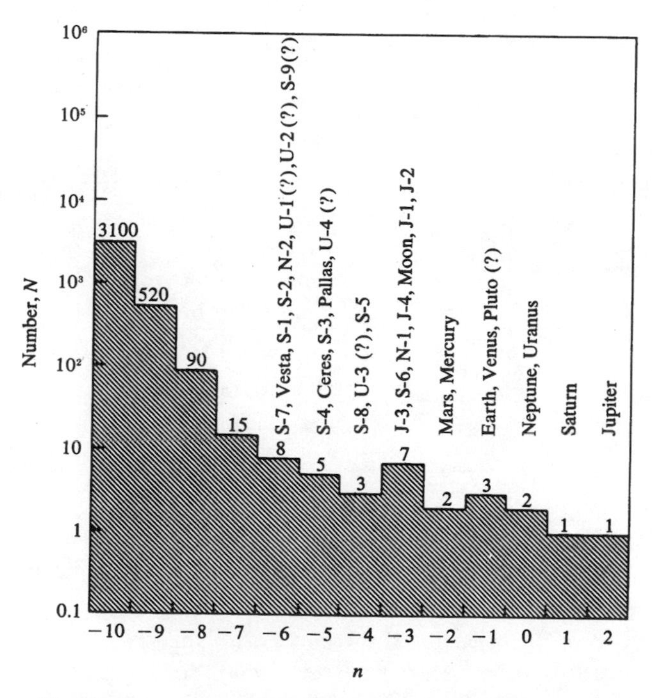

Figure 10. Number of objects in the solar system in the mass interval 3×10^n to $3 \times 10^{n+1}$ Earth masses.

Gas Capture and Retention. From the lower left-hand portion of Figure 7 (see page 28), it may be seen that all compact planetary bodies for which we have any data lie between two limiting curves. Those that lie on, or close to, the lower curve are spoken of as terrestrial planets and are rocky bodies with light atmospheres or no sensible atmospheres at all. Those that lie close to the upper curve must be bodies composed principally of compressed hydrogen and helium. Those that lie well between these limits must be bodies with very extensive gaseous envelopes. The apparent relationship between mean density and mass is given in Figure 11.

Until more is learned about the behavior of matter under extreme conditions of pressure, reliable quantitative estimates of the internal composition of these massive bodies cannot be made, although, if some assumptions are made regarding their modes of formation, some limits on their internal constitutions may be inferred. Much depends on the temperature conditions assumed at the time of formation and on the present critical temperatures at the level in the planetary atmosphere from which gases can escape.

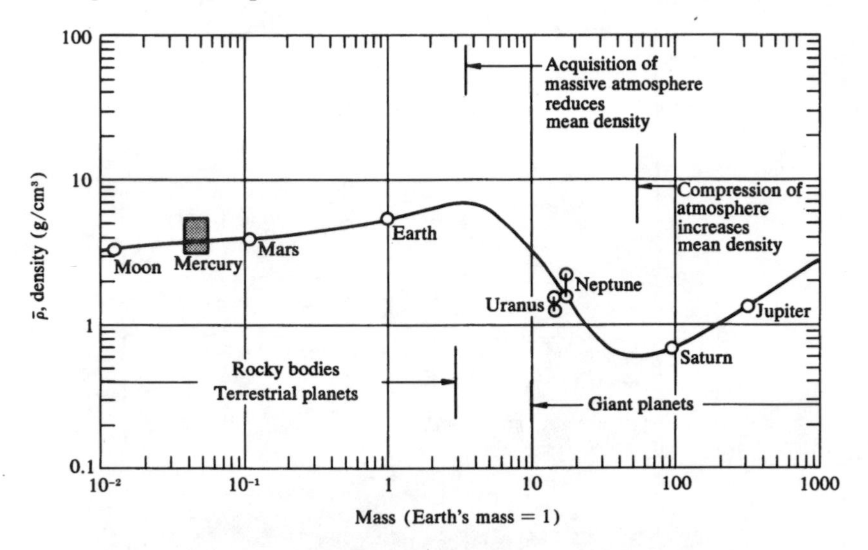

Figure 11. Mass-density relationship of the solar-system planets.

A number of theories on the escape of planetary atmospheres have been developed in recent years by Urey (1959), Spitzer (1952), and others. Unfortunately the more complex of these theories relating to the escape of planetary atmospheres require a detailed prior knowledge of the vertical temperature structure of the atmosphere, as well as a prior knowledge of the atmospheric composition; thus they can not be applied conveniently to generalized cases.

A more generally useful, although rough, yardstick to indicate the behavior of atmospheric gases was derived by Jeans (1916) and modified by Jones (1923):

$$t = \frac{\bar{v}^3}{2g^2R} e^{(3gR/\bar{v}^2)},$$

where t, the time in seconds required for the abundance of an atmospheric constituent to decrease to 0.368 (equal to $1/e$) of its former value, is related to the root-mean-square (rms) velocity of the molecules $\bar{v}$ (centimeters per second), the planetary radius R (centimeters), and the surface gravity g (centimeters per second per second).

This relationship may also be expressed as

$$t = \frac{k\bar{v}^3 R}{V_e^4} e^{(3V_e^2/2\bar{v}^2)},$$

where V_e is the velocity of escape from the planet and k is a constant. The exponential is the dominant term in this expression and can serve to differentiate between cases in which important gases may be captured and cases in which these gases may not be retained.

Specifically, if the root-mean-square velocity of the gas in question at the appropriate temperature *equals* the escape velocity of a planet, many of the gas molecules will have instantaneous velocities two or three times as high as the root-mean-square velocity and will also be moving in an upward direction. Consequently, the gas will escape rapidly. Similarly, if the root-mean-square velocity is one-half the planetary escape velocity, escape of the gas will still be rapid. Table 5 shows roughly the relationship

Table 5. Capture and Retention of Gases by Massive Bodies

$\frac{\text{Escape velocity}}{\text{Root-mean-square velocity}}$	e^{-1} life of atmosphere
1	0
2	0
3	a few weeks
4	several thousand years
5	~ 100 million years
6	~ infinite (permanent atmosphere)

between escape velocity and root-mean-square velocity and the ability of a body to capture and retain a gas. If a planet is to be able to capture a gas, the planetary escape velocity must be three or four times the root-mean-square velocity; for a planet to retain a gas permanently, the escape velocity must be about five or six times the root-mean-square velocity.

In capturing a gas, then, the important parameters are the root-mean-square velocity at the appropriate temperature, the escape velocity of the planet, and the rate of gas arrival. At any loss-rate level, if the gas is arriving at a rate substantially higher than the loss rate, an atmosphere will accumulate and the planetary mass will increase. Thus the escape velocity will tend to increase, and, moreover, the radius will increase rapidly as gas accumulates. These tendencies will aid in the accumulation of still more gas (and dust), and the gas-capturing process will accelerate until it is terminated by a lack of surrounding matter from which it can grow. The capture of gases as such, as opposed to gases dissolved in, or absorbed on, solid particles, probably does not play an important role in the development of the atmospheres of terrestrial planets. Gas capture on a grand scale would result in the formation of a giant planet.

With respect to the capture of gas by a massive, rocky, airless planet, the most important gaseous constituent for starting the "snowballing" process would be helium, for it is abundant and is more easily captured than hydrogen at the same equilibrium temperature. Once a certain amount of helium has been captured, hydrogen capture can proceed rapidly and a giant planet such as Jupiter or Neptune is the result.

The very tenuous upper region of a planetary atmosphere, where the gas density is so low that mean free paths of atoms or molecules are of the order of several kilometers or longer, is usually termed the "exosphere." In the exosphere there is a critical escape level that has been defined as the level at which there is only one mean free path for a very fast molecule moving vertically away from the planet (Urey, 1959). A dominating factor in the capture and retention of atmospheric constituents is the temperature at the critical escape level. For the Earth, the critical escape layer is apparently at an altitude of about 600 kilometers, and the temperature at this altitude is quite variable because of the changing intensity of solar radiation. Estimated temperatures fall within the range of 1000°K to 2000°K (CIRA-1961).

According to Chamberlain (1962), "the principal source of the thermospheric heat is probably far-ultraviolet radiation from the sun . . .. Other possibilities for augmenting this supply have been proposed from time to time, but their importance is still not demonstrated. Nevertheless, the problem of predicting the highest thermospheric temperatures on earth from purely theoretical considerations (or, if you prefer, the problem of explaining the observed temperatures) has not been completely and satisfactorily solved."

To illustrate the importance of temperature at the critical escape level, some exosphere temperatures have been *assumed* for the planets of our solar system and are shown in Figure 12. For purposes of illustration, it was assumed that exosphere temperature varies inversely with the perihelion distance (with Earth's maximum exosphere temperature taken as 2000°K). If this diagram were a true representation incorporating accurate maximum exosphere temperatures, some indication of major atmospheric constituents could be inferred, with gases below the points being retained and those above the points tending to escape. However, too rigid an interpretation should not be placed on the location of the points because of the extreme oversimplification of the problem and because the relationship between exosphere temperature and distance from the Sun is not known. The principal use of such a diagram is to indicate, in a general way, the interrelationships between gas retention, exosphere temperature, and intrinsic planetary properties. As mentioned earlier, some atmospheric

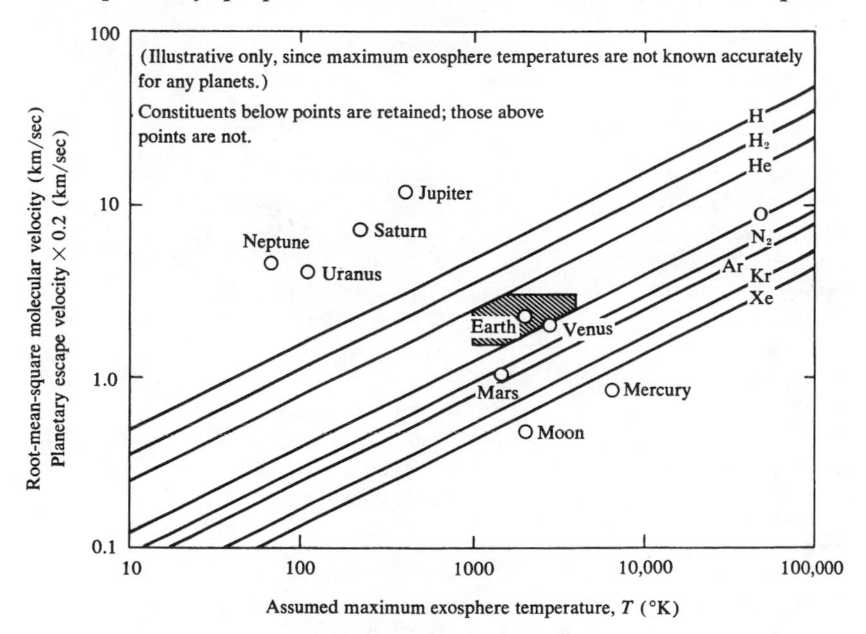

Figure 12. Planetary atmospheres.

constituents, although they escape slowly, may be replenished at the same rate by such processes as volcanism, photolysis, or radioactive decay. Consequently, these constituents may be present in dynamic equilibrium in a planetary atmosphere. The principal potential constituents of planetary atmospheres (in order of their molecular weight) are listed in Table 6. Of these gases, the most abundant (on the basis of universal elementary abundances) should be H, H_2 and its compounds; He; O, O_2 and its compounds; N, N_2 and its compounds; the compounds of carbon; Ne; Ar; and the compounds of sulfur. Krypton and xenon are relatively rare elements.

The compounds of nitrogen (except N_2), the compounds of sulfur, and CO_2 are quite water soluble and are active chemically; hence, they tend to be removed from atmospheres by surface reactions in the presence of water. Free oxygen is also very active chemically and would tend to disappear from an atmosphere unless it was continually replaced. In the presence of an excess of free hydrogen, the oxides would tend to be reduced to their corresponding hydrides.

Table 6. Potential Atmospheric Constituents

Constituent	Molecular weight	Constituent	Molecular weight
Atomic hydrogen, H	1	Nitric oxide, NO	30
Molecular hydrogen, H_2	2	Molecular oxygen, O_2	32
Helium, He	4	Hydrogen sulfide, H_2S	34.1
Atomic nitrogen, N	14	Argon, Ar	39.9
Atomic oxygen, O	16	Carbon dioxide, CO_2	44
Methane, CH_4	16	Nitrous oxide, N_2O	44
Ammonia, NH_3	17	Nitrogen dioxide, NO_2	46
Water vapor, H_2O	18	Ozone, O_3	48
Neon, Ne	20.2	Sulfur dioxide, SO_2	64.1
Molecular nitrogen, N_2	28	Sulfur trioxide, SO_3	80.1
Carbon monoxide, CO	28	Krypton, Kr	83.8
		Xenon, Xe	131.3

In the presence of an excess of free oxygen, H_2, CO, and CH_4 would tend to become oxidized to H_2O and CO_2.

The gases H_2, H_2O, O_2, and CO_2 may be dissociated photolytically to form H, H^+, OH^-, O, O^-, O_3 (transiently), and CO. Nitrogen (N_2) and CO are extremely stable molecules and are not readily dissociated by sunlight.

Water is a special case because of its high freezing point (triple point). It can be protected from photolytic dissociation by overlying gases (O, O_2, and O_3, for example) that absorb in the ultraviolet region of the spectrum. Low temperatures in the atmosphere which permit the freezing or condensation and precipitation of water can prevent it from rising to too high an altitude.

The most interesting planets from the human point of view would be those having the right combination of mass, radius, and exosphere temperature so that they could retain atomic oxygen, but not capture helium and thus "snowball." Further requirements would be suitable surface temperatures, tolerable gravitational forces at the surface (less than about 1.5 times the acceleration of gravity at the Earth's surface), a suitable partial pressure of atmospheric oxygen at the planetary surface (which implies photosynthesis), and the presence of both liquid water and permanent land areas.

Under the tentative assumptions made here, the planetary properties suggested by the diagonally shaded area of Figure 12 would be approximately those of habitable planets.

Classification of Planets by Atmospheric Characteristics. In general planets may be classified as follows: *Airless bodies* (more or less spherical aggregates having no sensible atmospheres)—for example, the Moon and Mercury; *planets with light atmospheres*—for example, the Earth, Mars, and Venus; *planets with massive atmospheres* (mainly hydrogen and helium)—for example, Jupiter, Saturn, Uranus, and Neptune.

Airless bodies (those lying below the Xe line in Figure 12) will conform more or less to the radius-density relationship of Equation (1), page 29.

Planets with light atmospheres (those lying between the lines for He and Xe in Figure 12) will also conform to the relationship of Equation (1), since their atmospheres will be small in mass relative to the mass of the rocky body. In general, the closer the body lies to the He line, the more massive its atmosphere will be. The position of any point representing a planet with a light atmosphere in Figure 12 gives an approximate clue as to its atmospheric composition. Gaseous constituents below the point will be retained, if present; those above will be lost. There also may be constituents in dynamic equilibrium, however, that are replaced as rapidly as they are lost, either by escaping to space or by entering into chemical combination.

Planets with massive atmospheres (those lying above the line for H or H_2) will have a composite structure consisting of a compressed rocky core, possibly a shell of water-ice; a compressed inner shell of metallic hydrogen (for very massive planets); and an outer gaseous atmosphere of hydrogen and helium, plus relatively minor quantities of other gases. Mean planetary densities will generally be less than 2.5 grams per cubic centimeter and total masses will be less than 10^3 or 10^4 Earth masses (the transition region between planets and stars).

Many of the details of the formation of planetary atmospheres are still somewhat conjectural. For the Earth, and possibly for most planets with light atmospheres, volcanism has apparently provided the primary ingredients from which the present atmosphere (and oceans) have developed.

According to Macdonald (1961), water is the chief component of volcanic gases at the Earth's surface, generally constituting more than 75 per cent, by volume, of all gases collected at volcanic vents. Other common constituents include carbon dioxide (CO_2), carbon monoxide (CO), sulfur dioxide (SO_2), sulfur vapor (S), sulfur trioxide (SO_3), hydrogen sulfide (H_2S), hydrogen chloride (HCl), ammonium chloride (NH_4Cl), and, in lesser abundance, hydrogen (H_2), hydrogen fluoride (HF), boric acid (H_2BO_3), methane (CH_4), nitrogen (N_2), and argon (Ar). Not all of these components are always present and their relative abundances vary considerably, but water is nearly always predominant and is often overwhelmingly so. It is probable that throughout the geologic ages, all the water on the Earth's surface has been produced by volcanoes. The principal volcanic gases have accumulated over the several billion years since the Earth formed, and a number of important physical and chemical changes have taken place. In the presence of water, carbon dioxide has been removed from the atmosphere and converted into carbonate rocks; the other water-soluble, chemically active materials have been dissolved and converted into various minerals; carbon monoxide and methane have been oxidized; hydrogen has been oxidized or has escaped; nitrogen and argon have remained in the atmosphere; oxygen has been produced through photosynthesis; and water has been retained on the surface as liquid and solid. If volcanism is accepted as the primary mechanism for the production of atmospheres of planets that are not massive enough to capture hydrogen and helium in large quantities, then an understanding of the relationship between degree of volcanic activity and planetary mass is essential for an understanding of the general course of atmospheric evolution.

Actually, our knowledge of the natural forces responsible for volcanic activity, earthquakes, and mountain formation on the Earth is still quite incomplete, so it is difficult at this time to specify general relationships between planetary mass and volcanism. One view, described by Bullard (1954), says that earthquakes and volcanic heat result from the mechanical energy associated with the distortion of the crust. Distortions of the crust would be produced by a general increase in the temperature of the interior of the Earth, with expansion, or by a general cooling, with contraction. Bullard comments that the present state of the Earth's surface strongly suggests contraction. The high interior temperatures of the Earth apparently are due to both gravitational compression and the accumulation of heat released by radioactive materials, principally uranium, thorium, and potassium-40, with radioactivity being the major contributor.

A planet smaller than the Earth would tend to accumulate less heat through gravitational compression during the period of formation. Its ratio of surface area to mass would be larger; hence, for a given internal temperature and thermal conductivity of the rocks, it could lose heat more rapidly. Small planets might also tend to have less concentration of metals toward the center, making the thermal conductivity of the outer parts greater. Thus small planets should have lower internal temperatures than large ones.

Planets with dense atmospheres and strong surface winds, particularly those with atmospheric constituents that can undergo a change of state at the prevailing temperatures (for example, water) will tend to have extensive erosion and horizontal transport of exposed crustal rocks. This tends to upset isostatic equilibrium in the crust and may tend to enhance volcanic activity.

Much more must be learned about the causes of volcanism and about the processes involved in the production of volcanic gases before it can be said that we have a complete understanding of the evolution and development of planetary atmospheres. There are still some little-understood factors relating to the Earth's atmosphere—its structure, composition, density gradient and temperature-altitude relationship, and circulation. But knowledge is being accumulated rapidly and undoubtedly much of this will be applicable to the more general problems of planetary meteorology.

Oblateness of Rotating Planets. Most of the discussion up to this point has been concerned with the properties of massive bodies that are not rotating strongly (see Figure 9). The rate of rotation, however, is an important intrinsic property of a planetary object, affecting its shape, surface gravity, and habitability. In a consideration of general planetary properties, the effects of rotation can not be ignored.

The shape of a rotating body isolated in space depends on its rate of rotation, its mean density, and the distribution of mass within the body. As a given body is caused to rotate more and more rapidly, its equatorial dimensions increase, producing a spheroidal or a pseudospheroidal shape.

Mathematical analyses have been made of the relationship between oblateness (degree of polar flattening) and density parameters for certain simple models. For example, rotating bodies of uniform density throughout (incompressible liquids) with one axis of symmetry (through the poles) are known as Maclaurin's spheroids; some of their properties have been derived by Lamb, Darwin, and Thomson and Tait (Jeans, 1929). Rotating bodies with all the mass concentrated at a point at the center of the body (Roche's models)* have been studied by Jeans, Poincaré, and others. Profiles of these idealized bodies (sectioned along the axis of rotation) are

* Roche's models have a nucleus of finite mass but infinitesimal volume, surrounded by an atmosphere of infinitesimal mass but finite volume (Roche, 1873).

shown in Figure 13. All real planetary bodies probably will lie between these two limits.

It is interesting to compare data for the planets of the solar system (Table 7) with the theoretical results mentioned above. This comparison

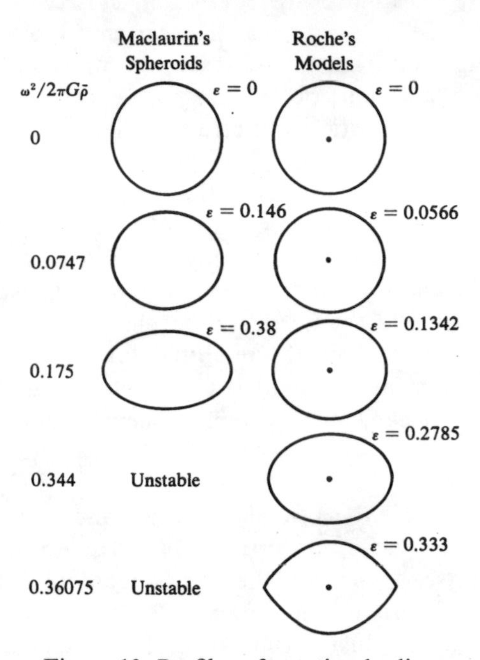

Figure 13. Profiles of rotating bodies.

is shown in Figure 14, in which oblateness ε (sometimes called ellipticity) is plotted against the parameter $\omega^2/2\pi G\bar{\rho}$, where ω is the angular velocity of rotation, G is the universal constant of gravitation, and $\bar{\rho}$ is the body's mean density. Oblateness ε is defined as $\varepsilon = (a - b)/a$, where a is the equatorial radius and b is the polar radius. This is to be distinguished from eccentricity, which is defined as $e^2 = (a^2 - b^2)/a^2$. The parameter ε is preferred here because it is more commonly used to describe the figures of planets. For Maclaurin's spheroids, axial symmetry is lost when $\omega^2/2\pi G\bar{\rho}$ exceeds 0.18712; for Roche's models, mass is lost at the equator when $\omega^2/2\pi G\bar{\rho}$ becomes larger than 0.36075.

In Figure 14, the points for Jupiter, Saturn, Uranus, and Neptune are observational and are not very definite because neither the densities nor the shapes are known with precision. This is especially true for Uranus and Neptune. Blanco and McCuskey (1961) have stated that for Uranus, a dynamical study yields $\varepsilon = \frac{1}{20}$, whereas direct observation yields

Table 7. Dynamical Oblateness of Rotating Bodies

Body	Period of rotation	ω^2 (rad²/sec²)	Mean density, $\bar{\rho}$ (g/cm³)	Oblateness, ε	$\omega^2/2\pi G\bar{\rho}$
Planets					
Mercury	87.97 days	6.83×10^{-13}	~4.24	0	~3.84×10^{-7}
Venus	(?)	(?)	5.32	0	(?)
Earth	23.935 hours	5.325×10^{-9}	5.52	0.00336	0.00230
Moon	27.32 days	7.09×10^{-12}	3.34	0	5.06×10^{-6}
Mars	24.623 hours	5.03×10^{-9}	4.0	0.0052	0.00300
Jupiter	9.842 hours	3.148×10^{-8}	1.34	0.062	0.056
Saturn	10.23 hours	2.91×10^{-8}	0.69	0.096	0.101
Uranus	10.82 hours	2.58×10^{-8}	1.26	0.05	0.049
			1.56	0.072	0.039
Neptune	15.67 hours	1.242×10^{-8}	1.61	0.02	0.0184
			2.27	0.0333	0.0292
	12.43 hours	1.97×10^{-8}	1.61	0.02	0.0130
			2.27	0.0333	0.0207
Maclaurin's spheroids					
$e = 0.1$	...	...	...	0.005	0.0027
0.2	...	...	...	0.020	0.0107
0.3	...	...	...	0.0455	0.0243
0.4	...	...	...	0.083	0.0436
0.5	...	...	...	0.134	0.0690
0.6	...	...	...	0.200	0.1007
0.7	...	...	...	0.2855	0.1387
0.8	...	...	...	0.400	0.1816
0.81267	...	...	...	0.417	0.18712
Roche's models					
	...	...	...	0.001	0.001333
	...	...	...	0.00505	0.00675
	...	...	...	0.01548	0.0207
	...	...	...	0.0561	0.0747
	...	...	...	0.1330	0.175
	...	...	...	0.1910	0.246
	...	...	...	0.224	0.284
	...	...	...	0.279	0.334
Critical equipotential	...	...	...	0.33333	0.36075

$\varepsilon = \frac{1}{14}$. They continue, "In this case both determinations are based on difficult observations. For Neptune, no definite observational evidence of flattening exists, but the inner satellite shows a strong precession effect from which Jackson has derived the oblateness quoted [i.e., $\varepsilon = \frac{1}{30}$] . . .." The oblateness of the Earth (one part in 297 or 298) has been

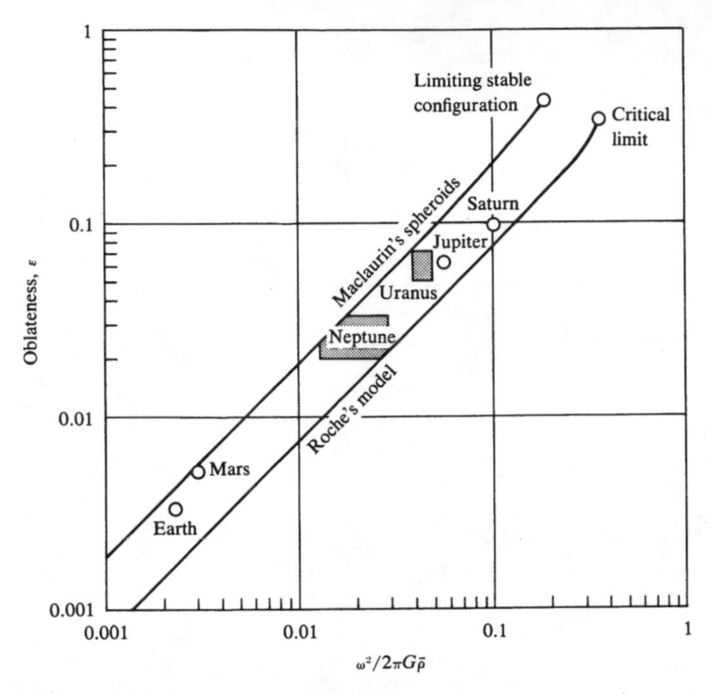

Figure 14. Oblateness of rotating planets versus angular-velocity–density parameter.

measured by direct survey and, independently, by observations of Earth satellites, while the oblateness or "dynamical flattening" of Mars (too small to observe directly with precision) was calculated by Struve, Burton, and Woolard (Urey, 1952) from changes in the orbital elements of Phobos and Deimos, the tiny satellites of Mars.

If the location of the points may be taken as correct, then their position with respect to the lines for Maclaurin's spheroids and Roche's model may be an indication of the degree of central concentration of mass within these bodies.

The surface gravity at various geocentric latitudes on these idealized bodies is shown in Figure 15. The surface gravity of real bodies, relative to the gravity when not rotating, would probably be intermediate between those of the Maclaurin and Roche models at the same value of $\omega^2/2\pi G\bar{\rho}$.

For the planets of our solar system, a very interesting relationship between mass and rotation rate strongly suggests that these are not

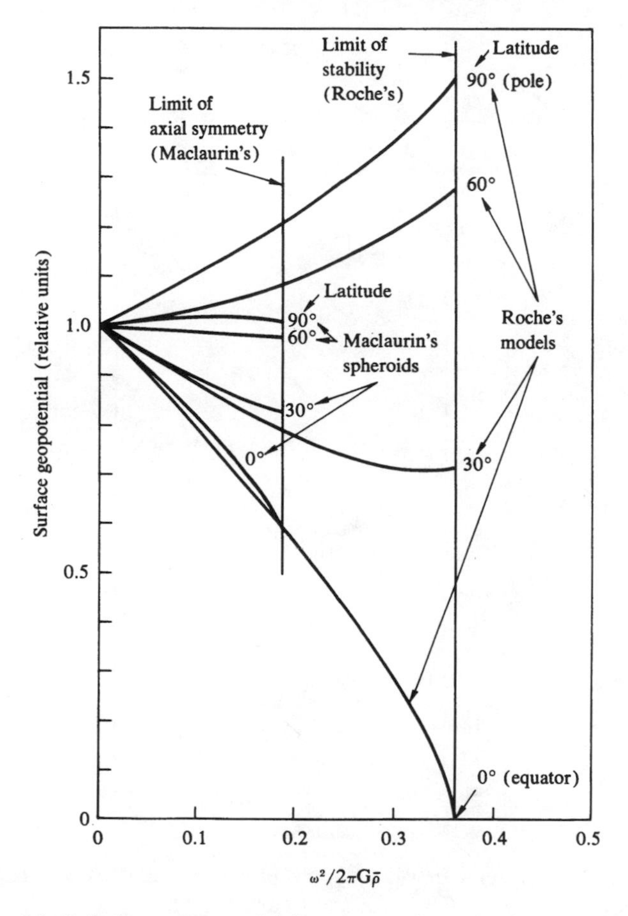

Figure 15. Relative geopotential (resultant of gravitational and centrifugal forces) at the surfaces of rotating bodies at selected geocentric latitudes.

independent of each other. Figure 16 shows the relationship between mass and rotational energy per unit mass for the Moon and all the major planets except Venus and Pluto (for which the rates of rotation are not known). The points for the Moon and Mercury fall well below the band in which the other points are found because their rotations have been stopped with respect to their primaries. The remaining points, however, are not too far from a line of slope +1, and this suggests (as an approximation) that the rotational energy per unit mass of a planet is directly proportional to the planet's mass M:

$$\frac{k_2}{2}\omega^2a^2 = \text{constant} \times M,$$

where k_2 is a dimensionless quantity related to the degree of central concentration of mass of a body. For a homogeneous body (Maclaurin's

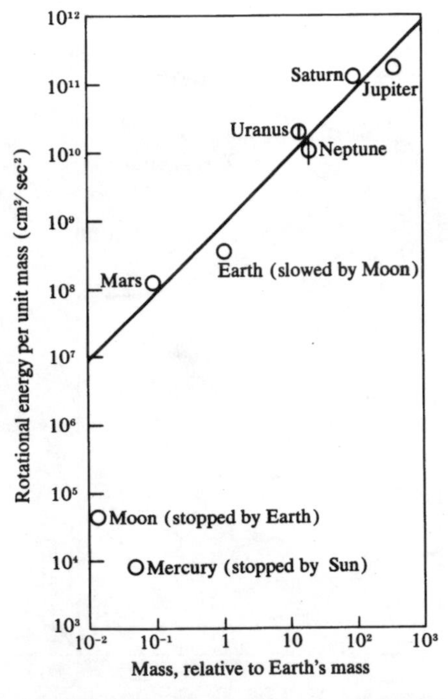

Figure 16. Rotational energy per unit mass versus mass (freely rotating planets of the solar system).

spheroids), $k_2 = 0.4$; for Roche's models, $k_2 = 0$. Values of k_2, ω, a, and M, used here, are given in the Appendix. (For the solar system, from the data for Mars, Uranus, Neptune, Saturn, and Jupiter, the constant in metric units is about 1.46×10^{-19} cm²/sec² g.) That the Earth falls below the curve is not surprising because we know that the Earth's rotation has been retarded by tidal effects due primarily to the Moon. If the Earth were to appear on the line, its period of rotation would be 15.4 hours, making the day about 64 per cent of its present length.

Atmospheric Density versus Altitude. The following equation gives an approximate relationship that shows how the density of a planetary atmosphere varies with altitude above the surface:

$$\rho_h = \rho_0 \, e^{-[gmh/KT]},$$

where ρ_h is the atmospheric density at a given altitude,
h is the altitude,
ρ_0 is the atmospheric density at the solid or liquid surface,
g is the attraction of gravity at the given altitude,
m is the mean molecular weight of the atmospheric gases,
K is the universal gas constant,
T is the absolute atmospheric temperature.

This equation states merely that atmospheric density, as a function of altitude, follows an exponential law and that a large planet (with a high surface gravity) should have a density gradient in its atmosphere that changes markedly with changes in altitude, while a small planet (with a low surface gravity) should have a more gradual density gradient. On the Earth, for example, the atmospheric density drops by a factor of about two for each 17,000 feet of ascent above the surface. On a larger planet having the same atmospheric molecular weight and temperature conditions, this halving of density would take place in altitude intervals of less than 17,000 feet. On a smaller planet, the halving of density would take place in altitude intervals of more than 17,000 feet. A large terrestrial planet might be said to have a "hard" atmosphere, and a small planet a "soft" atmosphere.

This relationship between size and density gradient has great bearing on the ease of entry of a space vehicle into a planet's atmosphere. Entry into the atmosphere of a small planet such as Mars can be achieved with less rigid restrictions on entry angle relative to the local horizontal and with lower g loadings on the passengers than entry into the atmosphere of larger planets such as the Earth.

The Composition of Planets. As we have seen, planets may be grouped into terrestrial planets and planets with massive atmospheres. Terrestrial planets consist almost entirely of rock minerals (chiefly the oxides of silicon, aluminum, iron, calcium, magnesium, sodium, and potassium) occurring in a great variety of crystalline types, mixtures, and compound forms. Below a certain critical mass range, the terrestrial planets must be composed almost entirely of nonvolatile constituents. Above this critical mass range, planets during their formative period can begin to capture helium and hydrogen along with the nonvolatile metallic oxides, silicates, *et cetera*. (The nonvolatile substances, however, are present in the universe to the extent of only about 1 per cent of the total.) Consequently, when a terrestrial planet grows to the critical mass and can begin capturing helium and hydrogen, its composition changes gradually from almost entirely nonvolatile material to principally hydrogen and helium.

Let us postulate a model universe composed of 99 per cent hydrogen and 1 per cent rock dust and assume, further, that a body cannot begin to capture gas effectively until its surface velocity of escape equals four times the root-mean-square velocity of hydrogen at 1500°K (assumed temperature at escape level), or $4.3 \times 4 = 17.2$ kilometers per second. This corresponds to a terrestrial planet of 3.2 Earth masses, which would be the critical mass M_C in this model. Let us assume that a body at this mass begins to capture all the materials that it encounters, gas and rock particles alike. From this point on, as it grows, its composition will be a simple function of the total mass M_T. The fraction that is gas, F_G, will be

$$F_G = \frac{0.99(M_T - M_C)}{M_T},$$

and the fraction that is rock, F_R, will be

$$F_R = \frac{M_C + 0.01(M_T - M_C)}{M_T}.$$

The resulting compositions are shown in Figure 17.

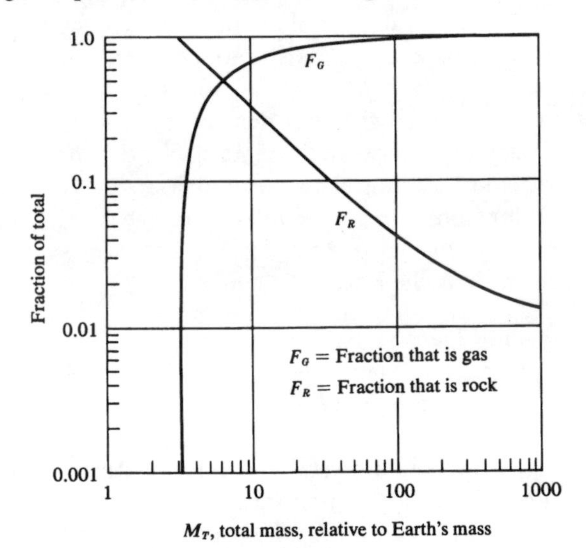

Figure 17. The changing composition of an accreting body after it becomes large enough to capture light gases (illustrative).

This is an all-or-nothing model as far as gas capture is concerned; accordingly, the gas fractions shown are probably quite high and the rock fractions may be rather low. Moreover, the chance capture of a few large terrestrial bodies by a massive planet could easily change the fractional composition by a significant amount. Figure 17 will serve, however, as an illustration of a body's transition from a terrestrial planet composed mainly of rock minerals to a giant planet composed mainly of gases.

Spacing of Planets in the Solar System. The regular spacing of the orbits of the planets of the solar system has interested students of astronomy ever since the planetary orbits were first established and measured with accuracy. Much significance has been attached to the Titius-Bode law of orbital distances.* The Titius-Bode law is not actually a law, but an empirical rule that fits quite well for the planetary semimajor axes out to and including Uranus (it breaks down for Neptune and Pluto). This rule fits, however, only if the asteroid Ceres is accepted as a major planet and if the odd mathematical discontinuity required to accommodate Mercury is included.

Another way of describing the planetary spacing would be to say that the orbits are spaced more or less regularly on a logarithmic scale. This is precisely what one would expect to find on the basis of orbital stability and the existence of forbidden and permitted regions in the restricted three-body problem (Dole, 1961). Limiting regions in the solar system are presented in Table 8 and Figure 18. In general, the concept may be summarized as follows: We consider each planet separately, starting with the largest (Jupiter). Each planet, together with the Sun, is considered as a member of a two-body system, having a certain mass ratio μ (mass of the smaller divided by the sum of the masses of the two), a certain orbital eccentricity e, and a certain mean distance $\bar{D}$ (semimajor axis of its orbital ellipse). First, consider the pair Jupiter–Sun. In its orbit around the Sun, Jupiter creates a broad annular band about 250 million miles wide centered roughly on its orbit; within this band, no small third body can exist in a stable orbit, and one would not expect to find a major planet growing by accretion within this band. The dimensions of this "forbidden" region are functions of μ, e, and $\bar{D}$.

Next, consider the Saturn–Sun pair. The forbidden region created by Saturn is an annular ring about 350 million miles wide which does nor overlap the forbidden region produced by Jupiter. It is as though

* One form of the Titius-Bode rule may be given as $r_P = 0.4 + 0.3(2)^x$, where $x = -\infty$ for $P = 1$ (Mercury) and $x = (P - 2)$ for $P \geq 2$, with P being the ordinal number of the planet counting from the Sun outward, and r_P being the mean planetary distance from the Sun in astronomical units.

Table 8. Limiting Regions for the Existence of Stable Near-circular Orbits as Applied to Planets of the Solar System

Pair	Mass ratio, μ	Semimajor axis of orbit, $\bar{D}$ (st. mi)	Eccentricity, e	Perihelion distance, D_m (st. mi)	Aphelion distance, D_M (st. mi)	r_{10} (st. mi)[a]	r_{20} (st. mi)[b]	r_{30} (st. mi)[c]
Sun–Mercury	1.39×10^{-7}	3.6×10^{7}	0.2056	2.86×10^{7}	4.34×10^{7}	2.83×10^{7}	53,500	4.39×10^{7}
Sun–Mars	3.23×10^{-7}	1.417×10^{8}	0.0934	1.286×10^{8}	1.550×10^{8}	1.267×10^{8}	295,500	1.572×10^{8}
Sun–Venus	2.45×10^{-6}	6.71×10^{7}	0.0068	6.65×10^{7}	6.75×10^{7}	6.465×10^{7}	304,000	6.95×10^{7}
Sun–Pluto	2.5×10^{-6}	3.67×10^{9}	0.2468	2.76×10^{9}	4.58×10^{9}	2.68×10^{9}	1.26×10^{7}	4.71×10^{9}
Sun–Earth	3.035×10^{-6}	9.29×10^{7}	0.0167	9.13×10^{7}	9.45×10^{7}	8.85×10^{7}	446,500	9.75×10^{7}
Sun–Uranus	4.36×10^{-5}	1.783×10^{9}	0.0472	1.693×10^{9}	1.866×10^{9}	1.567×10^{9}	1.98×10^{7}	2.02×10^{9}
Sun–Neptune	5.3×10^{-5}	2.793×10^{9}	0.0086	2.762×10^{9}	2.816×10^{9}	2.56×10^{9}	3.425×10^{7}	3.06×10^{9}
Sun–Saturn	2.86×10^{-4}	8.86×10^{8}	0.0557	8.36×10^{8}	9.35×10^{8}	7.29×10^{8}	1.84×10^{7}	1.082×10^{9}
Sun–Jupiter	9.55×10^{-4}	4.83×10^{8}	0.0484	4.59×10^{8}	5.06×10^{8}	3.76×10^{8}	1.492×10^{7}	6.24×10^{8}
Earth–Moon	1.23×10^{-2}	2.39×10^{5}	0.0549	2.257×10^{5}	2.52×10^{5}	1.443×10^{5}	16,460	4.08×10^{5}

[a] Maximum orbital radius of direct inferior planets.
[b] Maximum orbital radius of direct satellites.
[c] Minimum orbital radius of direct superior planets.

each planet produces a standoff distance that is a function of its mass and its distance from the Sun, and within which other planets cannot orbit in a stable manner. This holds true for Neptune, Uranus, the Earth, Venus, Mars, and Mercury, as well as for Jupiter and Saturn. None of the forbidden regions of these planets overlaps any other. The only exception is Pluto. Although Pluto's mean distance lies outside Neptune's forbidden

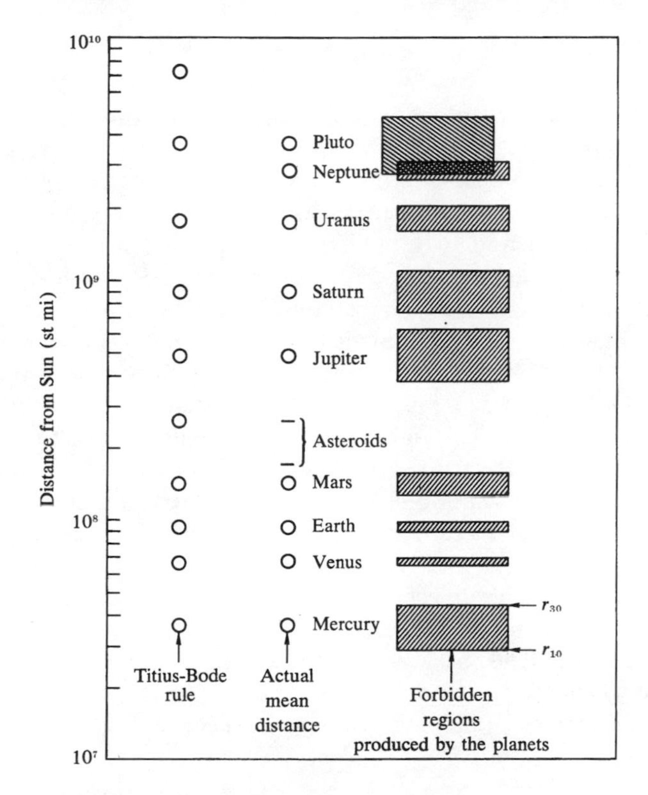

Figure 18. The spacing of planets in the solar system.

region, Pluto's orbit crosses into it; this may foreshadow an eventual catastrophic perturbation of Pluto's orbit by Neptune sometime in the distant future.

There are also wide gaps between certain adjacent forbidden regions within which small bodies can exist in stable orbits, notably the asteroid belt between Jupiter and Mars. Interestingly wide gaps are also present between the forbidden regions of Uranus and Saturn and between those of Neptune and Uranus, where small orbiting objects (as yet undiscovered) may well exist in large numbers.

This pattern of regularity of the solar system should also be found in other planetary systems. Forbidden regions take up about 50 per cent of our solar system; and if this is typical of other planetary systems (or multiple star systems), then it would be a simple matter to design any number of stable planetary systems by a random mechanical process.

OTHER PROBLEMS

Some of the simpler and more fundamental general properties of planets and the apparent relationships among these properties have been discussed in this chapter. A number of other interesting but more complex relationships among planetary parameters have not been covered in this very incomplete introduction to general planetology. Some of these include planetary effects due to the type of primary (or primaries in binary star systems); planetary surface temperature as a function of atmospheric properties, latitude, and orbital distance; temperature variability as a function of orbital eccentricity, rate of rotation, inclination of equator to plane of orbit, and position in binary star systems; tidal effects due to primary or satellites; planetary properties compatible with the existence of life forms; and the relative "sphericity" of massive aggregates. Other problem areas that have not been discussed include estimates of the number of planetary systems; evolutionary features of planets and stars (as related to general planetology); and the internal structure of planets.

A number of these subjects, insofar as they relate specifically to the question of habitability, will be discussed in subsequent chapters. Many of the basic problems of general planetology, however, will probably remain unresolved until more definitive observational data become available through scientific investigations in space.

CHAPTER 4

The Astronomical Parameters

PLANETARY PROPERTIES

The requirements of a habitable planet, from the viewpoint of human beings, were discussed in quantitative terms in Chapter 2. It is now appropriate to examine the intrinsic and positional properties of a planet and how they affect its habitability. As mentioned in Chapter 3, the intrinsic properties are mass, rate of rotation, and age; the positional properties are distance from the primary, orbital eccentricity, inclination of equator, satellite relationships, and the properties of the star around which the planet is orbiting.

Mass. The most basic of planetary parameters is mass, since this determines in great part the magnitude of the surface gravity, the atmospheric composition and surface pressure, whether or not water is retained, the level of radioactivity at the surface, the topographical roughness, and the rate of volcanism, to name only some of the more important resultant properties affected.

For a nearly spherical, slowly rotating, terrestrial planet, the relationships between mass, radius, surface gravity, and velocity of escape are shown in Figure 9 (see page 31). Now it will be recalled that, to be considered habitable, a planet must have a surface gravity of less than 1.5 *g*. From Figure 9, it may be seen that this corresponds to a planet that has a mass of 2.35 Earth masses, a radius of 1.25 Earth radii, and an escape velocity of 15.3 kilometers per second. These represent upper limits on the size and mass of a habitable planet. It is assumed that other limitations due to increasing mass have not occurred first, such as the surface being completely covered with water, or the atmospheric density being so high as to produce oxygen toxicity or nitrogen narcosis, or the atmosphere being so opaque that sunlight can not reach the surface with intensity levels high enough for effective photosynthesis.

What about the lower limit on mass? This will be reached when the mass is too small to retain a breathable atmosphere on its surface. As we have seen, the minimum tolerable atmospheric pressure for human beings is 2.1 pounds per square inch (for pure oxygen). Some nitrogen, however, is necessary for plants (being converted from atmospheric nitrogen into usable form by nitrogen-fixing bacteria or by lightning); hence, for an atmosphere consisting of, say, 90 per cent oxygen and 10 per cent nitrogen, the minimal barometric surface pressure would be about 2.3 pounds per square inch or 0.156 atmospheres (atm).

If we assume that all habitable planets must have surface temperatures in the approximate neighborhood of those on the Earth, then presumably their exosphere temperatures will also be similar to those in the Earth's exosphere. To prevent atomic oxygen from escaping rapidly from the upper layers of its atmosphere, the planet's escape velocity must be of the order of five times the root-mean-square velocity of the oxygen atoms in the exosphere. This is shown in Figure 12 (see page 37). We do not yet fully understand all the factors that are involved in producing the extremely high temperature in the Earth's exosphere (apparently 1000°K to 2000°K). However, if we take as a rough approximation that maximum exosphere temperatures as low as 1000°K are not incompatible with the required surface conditions of a habitable planet, then the escape velocity of the smallest planet capable of retaining atomic oxygen may be as low as 6.25 kilometers per second (5×1.25). Going back to Figure 9, this may be seen to correspond to a planet having a mass of 0.195 Earth mass, a radius of 0.63 Earth radius, and a surface gravity of 0.49 *g*. Under the above assumptions, such a planet could theoretically hold an oxygen-rich atmosphere, but it would probably be much too small to produce one, as will be seen below.

The ability of a planet to retain atomic oxygen does not necessarily guarantee that it will have an atmosphere containing free oxygen at the surface to the extent necessary for it to be called habitable. There are several processes that must take place before a planet can have a breathable atmosphere. First, there must be some mechanism for the production of free oxygen. Second, there must be some mechanism for the accumulation of free oxygen in the atmosphere to a partial pressure (inspired) of at least 60 millimeters of mercury.

On the Earth the existence of free oxygen in the atmosphere can probably be attributed entirely to the photolysis of water by green plants.

But there are also a number of processes tending to remove free oxygen from the atmosphere. Oxygen is consumed by the oxidation of minerals in the crust during weathering, by the respiration of animals, and by the oxidation of plants when they die and decay. When a plant decays or is burned completely, it uses up just as much oxygen as it produced while it was growing. Therefore, if oxygen is to accumulate in an atmosphere to any great degree, there must be some way in which suboxidized organic matter is prevented from being oxidized completely. If plants or animals are somehow buried or submerged in sediments, then there is a net gain of oxygen to the atmosphere. This is apparently the process through which free oxygen has been accumulated in the Earth's atmosphere. The great mass of all the living plants on the Earth's surface is also balanced by a corresponding mass of free oxygen in the atmosphere.

The principal point here is that weathering, erosion, and burial of organic matter in alluvial silts seem to be the primary agencies whereby uncombined oxygen can accumulate to any large degree in a planetary atmosphere.

Intuitively one would expect that small planets would have a lower rate of burial of organic matter than the Earth, while large planets would have more burial because of stronger erosive forces and more surface water. Moreover, a large share of the photosynthesis taking place on the Earth is carried out in the oceans (Weiss, 1952). One would therefore expect that low sea-land ratios might be associated with lower proportions of free oxygen in the atmosphere than would be the case with high sea-land ratios.

Furthermore, one would expect ample quantities of nitrogen always to be present in a planet with an atmosphere derived primarily from the inert constituents contained in the gaseous emanations from volcanoes.

In addition to the above, the atmospheric pressure p on the surface of a planet depends both on the mass of atmosphere per unit surface area (m/A) and on the surface gravity g,

$$p = \left(\frac{m}{A}\right) g.$$

For example, on Earth, $p = 1013$ millibars (mb), $g = 981$ centimeters per second per second, and $(m/A) = 1033$ grams per square centimeter; on Mars, p is of the order of 94 millibars, g is about 376 centimeters per second per second, and therefore (m/A) is about 250 grams per square centimeter (Davis, 1961). The Earth's atmosphere contains an abundance of oxygen (nearly 21 per cent), while Mars' atmosphere apparently contains essentially none. The smallest habitable planet will probably have a mass somewhere between that of the Earth and that of Mars, with a surface pressure and an atmospheric concentration of oxygen somewhere between theirs, but with an inspired oxygen partial pressure of 60 millimeters of mercury. By interpolating on Figure 6 (see page 14), the required minimal atmosphere should contain about 16 per cent oxygen and have a surface pressure of about 8 pounds per square inch (551 millibars). Using this figure as a guide, it is possible to estimate the mass of the smallest habitable planet. First, if we assume that our hypothetical planet has the same mass of atmosphere per unit surface area as the Earth (or that atmospheric pressure is proportional to g), then we can calculate the surface gravity (and mass) required to produce an atmospheric pressure of 551 millibars. The planetary mass thus determined is 0.25 Earth mass, which we know to be *too low*, since it is not probable that such a small planet would have the same mass of atmosphere per unit surface area as the Earth. Second, we can assume that surface pressure is a direct function of planetary mass and can interpolate between the Earth and Mars, using them as data points (Figure 19). By this method, the required mass to produce $p = 551$

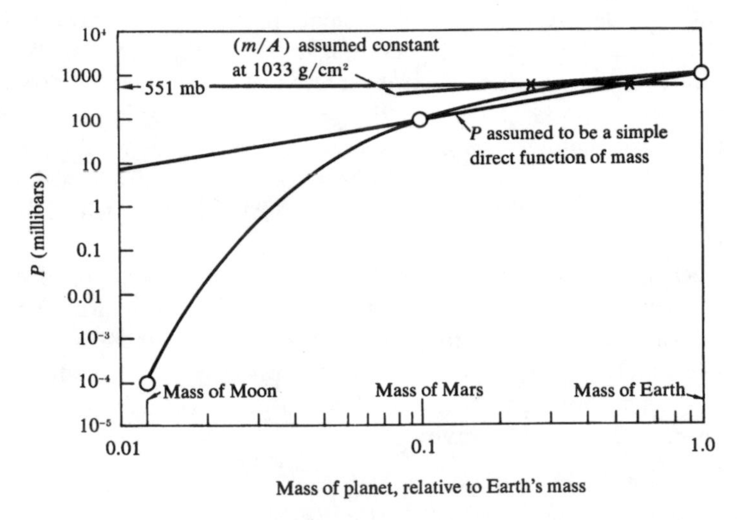

Figure 19. Atmospheric pressure at planetary surface as a function of planetary mass.

millibars is 0.57 Earth mass, which we know to be *too high* because the assumption that surface pressure is proportional to planetary mass gives an unreasonably high value for the Moon. With 0.25 being too low and 0.57 being too high, the appropriate value of mass for the smallest habitable planet must lie between these figures, somewhere in the vicinity of 0.4 Earth mass.

Since it is not possible to obtain a more precise determination of the minimum mass of a habitable planet, for our purposes the value of 0.4 Earth mass will be adopted as the lower limit of mass. This corresponds to a planet having a radius of 0.78 Earth radius and a surface gravity of 0.68 *g*.

For planets with light atmospheres, the density (or pressure) gradient in a planetary atmosphere depends on the surface gravity:

$$\rho_h = \rho_0 \, e^{-\left[\frac{g_0 m h}{KT}\left(\frac{R}{R+h}\right)^2\right]}$$

(for an isothermal atmosphere). To a fairly good approximation, $\rho_h = \rho_0 e^{-\alpha h}$. For h in feet, α for the Earth equals about 4.15×10^{-5} per foot. For the extreme examples of planetary mass discussed above (mass = 0.4

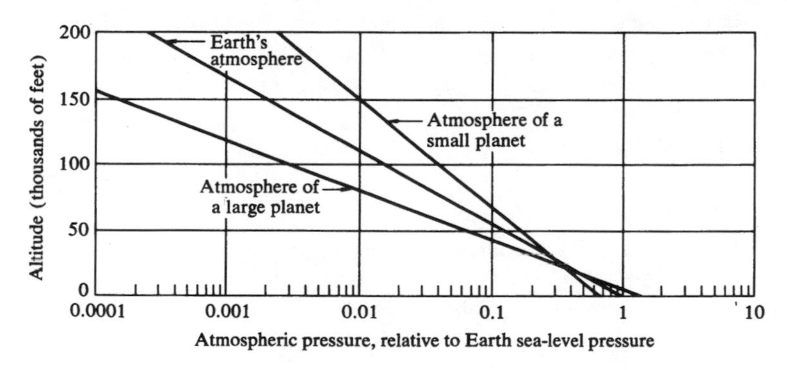

Figure 20. Atmospheric pressure as a function of altitude for large and small habitable planets compared with Earth (illustrative).

Earth mass, and mass = 2.35 Earth masses) where atmospheric molecular weight and temperature are the same as for the Earth, α would be 2.82×10^{-5} per foot and 6.23×10^{-5} per foot, respectively. The significance of this is seen more clearly in Figure 20; that is, small planets have "soft" atmospheres in which the atmospheric density changes slowly with altitude, and large planets have "hard" atmospheres in which the change in density with altitude is far more rapid. One effect of this is that at high altitudes (above, say, about 30,000 to 40,000 feet), the smaller planets should have denser atmospheres than the larger planets. This would have a distinct influence on such factors as the altitude ceilings of similar kinds of aircraft flying in the atmospheres of planets with different masses.

In the present study, then, it will be postulated that the mass of habitable planets may vary over the range 0.4 to 2.35 Earth masses; the radius may vary from 0.78 to 1.25 Earth radii; while surface gravity may range from 0.68 to 1.5 g. This size range is illustrated in Figure 21.

Within the given mass range, given proper temperatures in the atmosphere, the atmospheric composition and pressure would depend very much on the past history of volcanic activity and the exosphere temperature. Generally speaking, planets near the lower end of the permissible mass range might be expected to have developed lower internal temperatures during their period of formation and from subsequent radioactivity, to have cooled more rapidly and to have thicker crusts. They might also be expected to show less volcanic activity and, consequently, to have less atmospheric gas and lower atmospheric pressures at their surfaces. Those near the upper end of the permissible mass range might exhibit more volcanism and have higher atmospheric pressures.

Within the above mass range, the planets of larger mass possibly would tend to have developed more internal classification or stratification of minerals and elementary materials; that is, there might be a greater concentration of dense materials at their centers, leaving their crusts relatively less rich in certain heavy metals and heavy minerals.

Rate of Rotation. As discussed in Chapter 3, the rate of rotation of a planet, together with its mean density and degree of concentration of mass toward its center, uniquely determine its oblateness. However, at high rotation rates the internal mass distribution may be affected by the rotation, so the two are not completely independent. Other characteristics also depend in part on rotation rate—surface gravity as a function of latitude, the daily temperature cycles, atmospheric circulation patterns and wind velocities, and, possibly, the magnetic field.

In general, the more rapid the rotation rate, the smaller the day-to-night temperature differences would be. The meteorological factors (wind velocities, cyclone patterns, *et cetera*) are extremely difficult to estimate in a quantitative manner because not enough is known at present about general planetary meteorology or the prediction of climates from astronomical parameters.

From the standpoint of human habitation, there are two limits related to rotation rates. For slow rotation rates, a limit would be reached when daytime temperatures became excessively high in the low latitudes below some critical latitude and when nighttime temperatures became excessively low poleward from this same latitude, or when the light-darkness cycle became too slow to enable plants to live through the long hot days and long cold nights. If rotation rate were increased steadily, a limiting point would be reached when surface gravity at the equator fell

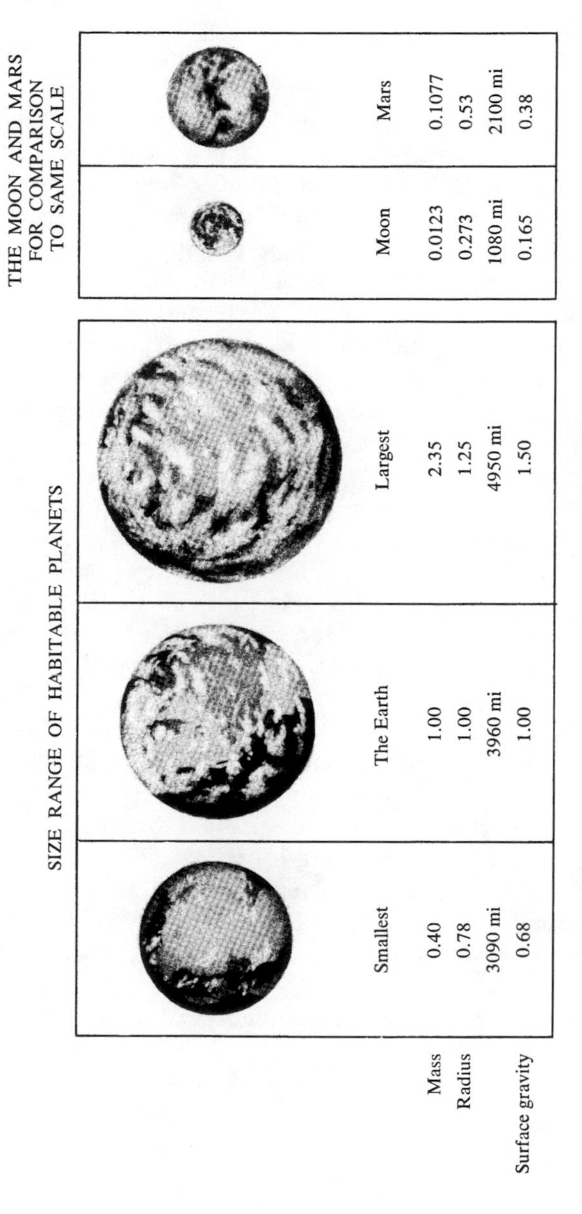

Figure 21. The extreme size range of habitable planets.

to zero and matter was lost from the planet, or when the shape of the surface became unstable and axial symmetry was lost.

Just what extremes of rotation rate are compatible with habitability is difficult to say. These extremes, however, might be estimated at, say, 96 hours (4 Earth days) per revolution at the lower end of the scale and 2 to 3 hours per revolution at the upper end, or at angular velocities where the shape becomes unstable because of the high rotation rate.

A special case, that in which the planet's sidereal day is precisely equal to its year, so that one side of the planet perpetually faces the primary while the other side is in perpetual darkness, might be thought to be compatible with habitability—that is, temperatures near the terminator (day-night line) might be in the desired range. In our solar system, Mercury fits into this category as far as rotation rate is concerned. Just what would happen to a planet's atmosphere under these circumstances is a somewhat debatable question. Would the atmospheric circulation be strong enough to prevent all the gases from condensing on the dark side? Or would all the water and carbon dioxide, at least, precipitate out in the extreme cold of the dark side? If it is assumed, as seems reasonable, that the day-equals-year situation did not come into being *ab initio* but was preceded by a long slowing-down period, then, during this time, all of the planet's water might well have been lost by photodecomposition with the subsequent escape of hydrogen. The day-equals-year case then may be ruled out as indicating either that all the water is precipitated out in solid form on the dark side or that the planet is completely dry. In any case, there would be no oceans of liquid water on the planet, and, consequently, it would not be habitable according to the present definition of the term.

Considering high rotation rates, it is apparent that the force of gravity becomes a function of latitude, being lowest at the equator and higher in high latitudes. For an Earth-like planet with a 3-hour period of rotation, for example, if it is assumed that the mean density is unchanged (5.52 grams per cubic centimeter), the oblateness would be about 0.24 and the force of gravity at the equator about 0.7 *g*. The gravitational field at other latitudes would be higher and would depend on the internal density distribution.

From this it may be seen that surface gravity depends on rate of rotation as well as on mass, and this dependence may affect previous statements about the upper limit of mass of a habitable planet. Based on data on planets of the solar system, however, it seems probable that the rate of rotation is a rough function of mass, as shown in Chapter 3. The correlation is not perfect, but the trend displayed is probably not due to chance alone. Also, it is reasonable to hypothesize that, during the process of planetary formation (by accretion), each captured particle of mass affected rotational energy unless it fell squarely dead-center on the planet's disk. The net effect of capturing many particles has been an increase in rotation rate with increasing mass. Since all the planets of our solar system except Uranus rotate in the same direction as their orbital motion, there was evidently a tendency for the incoming particles to impact so as to impart a direct spin to the planets. This hypothesis is illustrated in Figure 22.

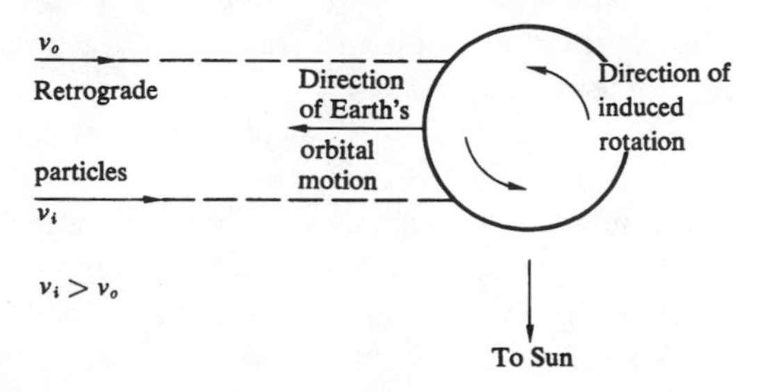

Figure 22. Rotation of the Earth produced by impacting particles.

The very low rotation rates exhibited by Mercury, the Moon, and probably Venus (although its true rate of rotation is not yet known) are apparently due to tidal braking effects. This subject will be discussed later.

Age. A certain amount of time must elapse before a newly formed planet can have surface conditions suitable for life. The sequence of events for an Earth-like planet might proceed according to the thirteen steps listed below.

1. A planet is formed by the gradual accretion and capture of small particles.

2. After the accretion process has been terminated because of a lack of growth materials, the surface is airless, or very nearly so.

3. The interior of the planet is extremely hot as a result of gravitational compression. Internal hydrostatic readjustments are taking place; denser materials, such as iron, are flowing slowly downward and lighter materials, such as certain silicates, are flowing upward. Because of the high viscosities involved, the internal readjustments take place over a long period of time. They also produce movements in surface materials, with extensive volcanism, crustal movements, or earthquakes. Localized heating of crustal rocks by friction and the decay of radioactive materials in the mantle produce high temperatures, and absorbed gases are released.

4. The lighter gases escape and the heavier gases are retained. Gases such as methane and ammonia may begin to accumulate if their rate of evolution exceeds their rate of escape. Photodissociation of water

vapor by solar radiation takes place, with the hydrogen escaping and the oxygen entering into chemical oxidation of surface materials.

5. Those stable gases that can be retained (nitrogen and carbon dioxide and, possibly, methane and ammonia) begin to accumulate and an atmosphere begins to form.

6. As the atmosphere becomes thicker and volcanic activity continues, a point is reached at which the rate of production of water vapor exceeds the rate of loss by photodissociation and traces of oxygen and ozone can appear in the atmosphere.

7. The oxygen and ozone absorb in the ultraviolet region of the solar spectrum (the wave lengths responsible for photodecomposition of water) and water vapor can now begin to accumulate more rapidly. The presence of even small amounts of ozone also produces a more stably stratified stratosphere, so that water vapor is unable to diffuse upward as rapidly. This is known as the atmospheric "cold trap" and is an important factor in the retention of water vapor.

8. When the concentration of water vapor in the atmosphere reaches the dew point or frost point, liquid water or frost condenses locally. When the atmospheric pressure has become high enough (greater than 6.3 millibars), the water condensed on the surface can appear as a liquid, provided that the temperature conditions are right (slightly above 0°C).

9. With continued volcanism, bodies of water appear on the planet and most of the carbon dioxide goes into solution, forming carbonic acid and reacting to form carbonate rocks. The ammonia also goes into solution and enters into reactions. Now the atmosphere consists mainly of nitrogen and methane, with water vapor as a variable constituent.

10. The bodies of water increase in size and join to form oceans. Rainfall becomes more prevalent; weathering begins to become significant; soluble minerals are washed into the oceans.

11. More complicated chemical compounds begin to accumulate in the oceans. Lightning discharges form small quantities of the oxides of nitrogen; these dissolve to form nitric acid and nitrates. Sulfur dioxide from volcanoes dissolves to form sulfuric acid and sulfates.

12. At some point, life appears, and eventually photosynthesis is established; free oxygen begins to accumulate in the atmosphere. [So much has been written in recent years about the origin of life on the Earth and elsewhere that no attempt will be made here to review these ideas.* It is significant that life, in the opinions of those who have studied the subject intensively (Edsall and Wyman, 1958; Henderson, 1958; Sidgwick, 1950),

---

* The interested reader is referred to Calvin (1955, 1959, 1961); Haldane (1928, 1954); Horowitz (1956, 1958); Oparin (1938, 1957, 1961); Tax (1960); and Wald (1954).

implies life based on the compounds of carbon. Sidgwick has summed up unequivocally, "Carbon is unique among the elements in the number and variety of the compounds which it can form. Over a quarter of a million compounds have already been isolated and described, but this gives a very imperfect idea of its powers, since it is the basis for all forms of living matter. Moreover, it is the only element which could occupy such a position. We know enough now to be sure that the idea of a world in which silicon should take the place of carbon as the basis for life is impossible; the silicon compounds have not the stability of those of carbon, and in particular it is not possible to form stable compounds with long chains of silicon atoms."]

13. After a long period of time, during which the prevalence of plants (or living forms carrying on photosynthesis) increases, the oxygen concentration of the atmosphere reaches the minimum value required by human beings; the volcanic activity level has slowed down, the meteorite-infall rate has diminished, and the planet may be considered habitable.

How long does this entire process take? One billion, 2 billion, 3 billion years? It is not possible to say with much accuracy, but the amount of time is surely of this order of magnitude. Thus, even though a planet has all the other essential attributes from an astronomical point of view, it must also be of a certain age before it can be considered habitable.

From the evolutionary point of view, or from the above sketchy chronological sequence, it may be seen that several factors could interfere with the development of suitable conditions on the surface of a planet. If the planetary mass were only slightly too small, the rate of juvenile water production by volcanic activity would be too low to balance the loss rate by photodissociation, and water would never accumulate on the surface. If the mean surface temperature were too high, water would never condense on the surface; instead, all of it would remain in the atmosphere, where it would continue to be lost by photodecomposition. No oceans would form, and carbon dioxide would become a major constituent of the atmosphere.

In general, it is probably safe to say that a planet must have existed for 2 or 3 billion years, under fairly steady conditions of solar radiation, before it has matured enough to be habitable.

Distance from Primary and Inclination of the Equator. The two parameters, distance and inclination, must be considered together because habitability depends on the two in combination, rather than on each independently. Orbital eccentricity is also interrelated with these parameters in determining habitability. As will be indicated later, however, the eccentricity requirements seem to be somewhat less restrictive than those of distance and inclination. In the present discussion it will be assumed that the orbital eccentricity is zero (the orbit being a circle centered on the primary).

Actually the radiation received by a planet from a particular star with given luminosity L may be expressed either in terms of the distance r or the illuminance E, because $E = L/r^2$. Illuminance is the term more generally useful in discussing the habitability of planets, so it will be employed in the remainder of this section. The terms are defined so that if r is in astronomical units and L is in terms of the Sun's luminosity, then E is in terms of the solar constant above the Earth's atmosphere at 1 astronomical unit (solar constant = 1.94 g cal/cm^2 min, or 1.35×10^6 erg/cm^2 sec).

In this connection, it is useful to introduce the term "ecosphere" (from the Greek, *oikos*, house, with the combining form *oiko-* denoting habitat or environment), a word apparently first used by Strughold (1955). For present purposes, ecosphere will be used to mean a region in space, in the vicinity of a star, in which suitable planets can have surface conditions compatible with the origin, evolution to complex forms, and continuous existence of land life *and* surface conditions suitable for human beings, along with the ecological complex on which they depend. The ecosphere lies between two spherical shells centered on the star. Inside the inner shell, illuminance levels are too high; outside the outer shell, they are too low.

Now it is a difficult problem to predict temperatures at a particular location on the surface of a planet as functions of illuminance and equatorial inclination. The problem becomes extremely complicated when a planet has atmospheric circulation and irregularly shaped ocean and dry-land areas. We do not even have acceptable theories about the causes of the glacial periods and of the dependence of climatic changes on the distribution of land and sea and the inclination of the equator. Attempts to calculate even mean annual temperatures on the Earth's surface on theoretical grounds have not been highly successful (Milankovitch, 1930). For this reason it was decided to use empirical methods here for the determination of planetary surface temperatures, using the Earth as a standard.

The empirical method employed assumed Earth-like planets with optically thin atmospheres and a cloud cover of approximately 45 per cent. Theoretical temperatures were calculated at various latitudes and seasons for rapidly rotating, nonconducting, black spheres (of various axial inclinations) that were half-illuminated by a distant point source. A relationship was then found between these theoretical temperatures and the actual observed temperatures on the Earth's surface (Figure 23), and this relationship was used as a basis for predicting mean surface temperatures on planets of any inclination at any latitude and at summer solstice, winter solstice, and the equinoxes.

Finally, habitability figures were computed by applying the rules that a region is habitable only if the mean annual temperature lies between

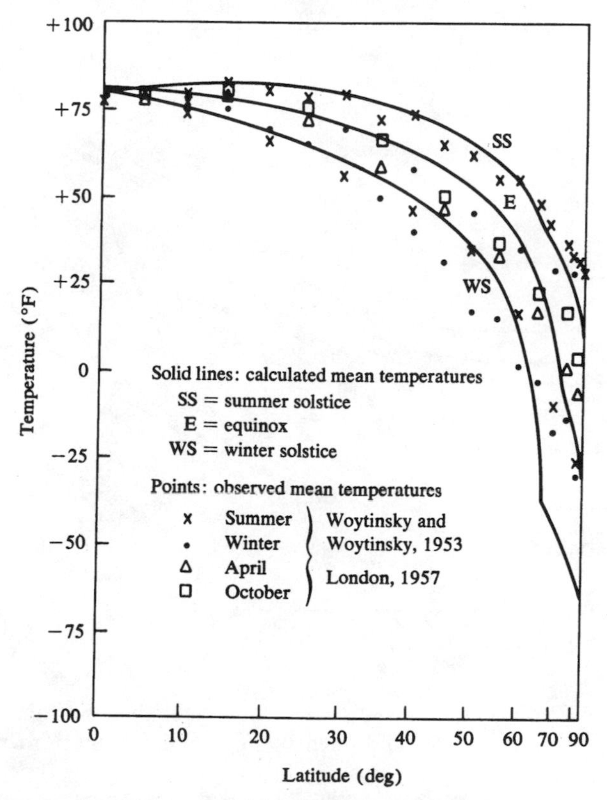

Figure 23. Comparison of calculated mean temperatures and observed mean temperatures for the Earth (equatorial inclination, 23.5°).

32°F and 86°F and if the highest mean daily temperature is less than 104°F and the lowest mean daily temperature is higher than 14°F. The results are summarized in Figure 24, which shows the percentage of planetary surface area that is habitable under the above definition as a function of illuminance and inclination.

If a planet must have 10 per cent of its surface in the proper climatic region to be considered habitable, then it may be seen that axial inclinations to approximately 80 degrees are tolerable, and that illuminance can vary from approximately 0.65 to 1.9 times Earth normal. These bounds circumscribe the "ecosphere" as applied to planets for human habitation. In our solar system, the ecosphere extends from 0.725 astronomical unit to 1.24 astronomical units, its inner edge reaching to the orbit of Venus (0.723 astronomical unit, mean distance) and its outer boundary reaching halfway to the orbit of Mars (1.526 astronomical units, mean distance).

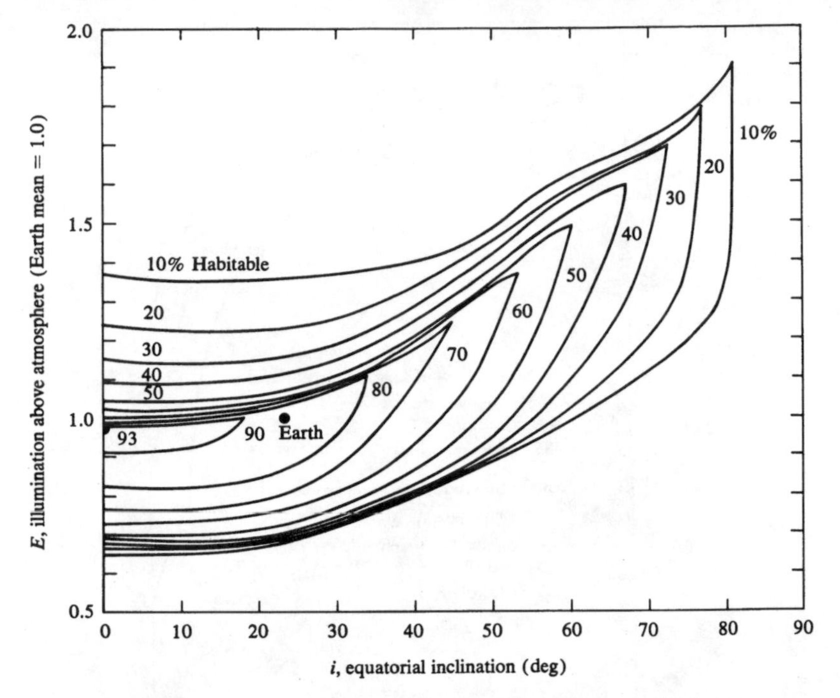

Figure 24. Percentage of surface area that is habitable as a function of inclination of equator and intensity of radiation from primary.

For small values of inclination, illuminance levels at the boundaries are 0.65 and 1.35 times the average value arriving at the top of the Earth's atmosphere.

So many assumptions had to be made in arriving at quantitative bounds of illuminance and inclination for habitable planets that the results should not be considered rigid and absolute. Rather they are used as a guide for making order-of-magnitude estimates of the prevalence of habitable planets in the Galaxy.

Orbital Eccentricity. For moderate values of orbital eccentricity, up to $e = 0.2$ or 0.3, the mean annual temperature of a planet seems to be virtually independent of eccentricity. The seasonal temperature cycle as a function of latitude, however, will depend on the relative positions of aphelion and winter solstice (for one hemisphere), as well as the absolute period of revolution of the planet. Planets with short "years" could remain habitable even with relatively high values of eccentricity because seasonal changes would tend to become blurred by the natural sluggishness of response, or temporal "lag," in seasonal change. Planets with long "years" could not tolerate such high values of orbital eccentricity.

Because of precession of the axis, there are, inevitably, times in the history of a planet when aphelion and winter solstice coincide, giving maximally cold winters and hot summers to one hemisphere but mild seasonal changes to the other. (Half a precession period later, then, the situation is reversed.) The rate of precession of the Earth's axis is 50.5 seconds per year (roughly one-third caused by the Sun and two-thirds by the Moon), so it takes about 25,600 years for one complete "wobble" of the axis due to precession. In general, the rate of precession of a planet's axis ψ is directly proportional to the mass of the primary M_1, the oblateness ε of the planet, and the cosine of the inclination of the equator i; it is inversely proportional to the angular velocity of rotation ω of the planet, and the cube of the semimajor axis of the planet's orbit r (G is the universal constant of gravitation). The relationship for the rate of precession may be given as follows (Lamb, 1929):

$$\psi = \frac{3}{2} G \frac{M_1 \varepsilon \cos i}{\omega r^3}.$$

High precession rates would be a characteristic of planets in close-in orbits having high spin rates (since ε is roughly proportional to ω^2) and low inclinations; low precession rates would be characteristic of planets in distant orbits having low spin rates and high inclinations.

Some trial calculations, using the Earth as an example and assuming that aphelion coincided with winter solstice, indicated that habitability is not affected in any significant manner by eccentricities up to 0.2. Because of the inevitable assumptions required, and since high values of eccentricity appear to be relatively improbable for bodies of planetary mass, it was considered unnecessary for present purposes to explore all combinations of eccentricity, inclination, period of revolution, and illuminance. Instead it was decided to take $e = 0.2$ as an arbitrary upper limit for the orbital eccentricity of habitable planets.

PROPERTIES OF THE PRIMARY

From previous discussions of the dependence of habitability on planetary age, it may be seen that the primary star must emit light and heat at a fairly constant rate for a period of at least 3 billion years. It may be appropriate at this point to give a very brief review of the classification of stars. Although all the stars (except the Sun) are so far away that they can not be seen as disks but merely as points of light, a great deal of information has been accumulated about them from measurements of their luminosities, distances from the Sun, and surface temperatures; from studies of the bright and dark (absorption) lines in their spectra; and from measurements made with various instruments such as the interferometer, which enables astronomers to measure the diameters of certain stars. All but a minor fraction of the stars belong to the main sequence; and, according to presently accepted views of stellar evolution, stars in the main sequence are in the stable phase of their existence and are converting hydrogen into helium at a steady rate. After they have consumed a certain fraction of their available hydrogen, stars leave the main sequence, expand greatly to become red giants, and then go through various rapid evolutionary phases before ending up as white dwarfs, perhaps after an explosive loss of mass, as may occur in the flaring up of a nova. In general, the evolution of a massive star proceeds more rapidly than that of a star having small mass.

Nearly all stellar spectra can be arranged in a sequence in which the intensities of the absorption lines change continuously. The spectral sequence contains seven main groups or classes designated (from hottest to coolest) as O, B, A, F, G, K, and M. The subdivisions of the groups are indicated by a number from 0 to 9 following the letter—for example, B0, B1, B2, . . ., B9, A0. The Sun is classified as G0 or, sometimes, as G2. Among main-sequence stars, the class O stars (very rare) are the most massive and the hottest (with surface temperatures up to 50,000°C). They have the largest diameters and the lowest densities, and they spend the least time in the main sequence. The class M stars (very abundant) ate the least massive and the coolest (with surface temperatures as low as 3000°C). They have the smallest diameters and the highest densities and they spend extremely long periods of time in the main sequence. The B, A, F, G, and K stars are intermediate in respect to these properties.

The only stars that conform with the requirement of stability for at least 3 billion years are main-sequence stars having a mass less than about 1.4 solar masses—spectral types F2 and smaller—although the relationship between mass and time of residence in the main sequence is probably not known with great accuracy and is subject to future revisions (see Figure 25). Stars having masses greater than 1.4 solar masses spend less than 3 billion years in the main sequence and then go into evolutionary phases in which the energy output changes rapidly.

There is also a limit at the low end of the mass range determined by the braking effect of the primary on the planet's rate of rotation. It is evident that low rates of rotation are incompatible with human habitability requirements and also that tidal effects can cause a planet's rotation rate to slow down until one side always faces the primary and there is no longer an alternation of day and night on the planetary surface.

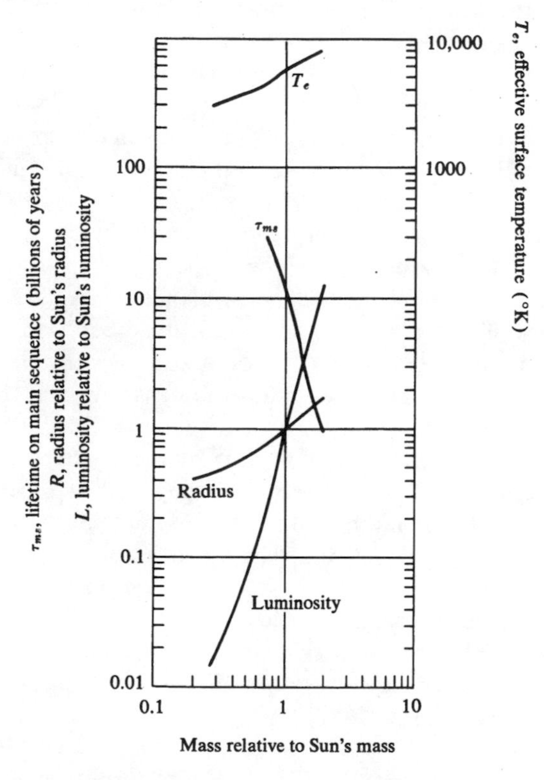

Figure 25. Some characteristics of the smaller main-sequence stars.

The tide-generating potential of an astronomical body may be shown to be proportional to

$$\frac{M_A R_B^{\,2}}{r^3},$$

and the maximum height h of equilibrium tides in deep water on the surface of an Earth-like planet may be shown to be proportional to

$$\frac{M_A R_B^{\,4}}{M_B r^3},$$

where M_A is the mass of the disturbing body, M_B is the mass of the disturbed body, R_B is the radius of the disturbed body, and r is the distance between the two bodies (Webster, 1925). If we let the maximum height of tides in deep water on the Earth due to the Moon equal unity when dimensions are expressed in feet, then $h = 0.85\ M_A R_B^4 / M_B r^3$. Actually, on the Earth, tides in the middle of the ocean are only a foot or so in height above mean sea level, but the fact that there is a finite quantity of water confined in more or less rigid connecting basins with highly irregular rim and bottom shapes results in a piling up of water to heights of many feet on shores and in bays. Also, strong tidal currents are produced along some coastlines and through certain channels and straits, and very complicated flows of water are generated on various parts of the Earth's surface, depending on local topographical relationships.

The height of tides alone would be important in our discussion of habitable planets in general, since all such planets must have more or less extensive bodies of water on their surface. But probably more important are the rotation-retarding effects of tides, a subject still only imperfectly understood, although it has been studied intensively for the past two hundred years. It is now widely accepted that the dissipation of energy by tidal friction in all of the shallow seas of the Earth is quantitatively adequate to account for the observed slowing of the Earth's rotation. Other factors may also be important, such as bodily tides in the Earth and changes in the Earth's moment of inertia due to secular changes within the Earth, changes in the oceans or the sea level, tides in the atmosphere, or interactions between the magnetic fields of Earth and Sun.

Jeffreys developed a theoretical relationship between heights of tides and tidal torques indicating that tidal rotation-retarding torques should be proportional to h^2 (Urey, 1952).

From Table 9 which gives calculated values of h and h^2 for various pairs of bodies of the solar system where the disturbed body is terrestrial in type, it may be seen that high values of h^2 always result in a stopping of the rotation of the disturbed body with respect to the disturbing body. The break point seems to come somewhere between $h^2 = 1.2$ and 2.0, since we know that the Earth's rotation has been retarded somewhat, principally by the action of the Moon, while the rotation of Venus is apparently quite slow, although its period of rotation is not known with any accuracy. It is quite probable that the retardations in the rotations of Mercury and Venus (assuming that Venus has no oceans) have been produced by bodily tides. Although some of the large satellites of the giant planets also produce high values of h^2 on their primaries, and may have produced some retardations in their rotations, the tides would be largely atmospheric in these cases and probably not nearly as effective in retarding

Table 9. Tidal Retardation Effects in the Solar System

Body A	Body B	Effect of body A on body B h	h^2	Rotation of B with respect to A
Sun	Mercury	3.11	9.65	stopped
Sun	Venus	1.33	1.77	very slow[a]
Sun	Earth	0.454	0.206	24 hours
Sun	Moon	0.204	0.0416	29.5 days
Sun	Mars	0.0976	0.00952	24.68 hours
Moon	Earth	1.00	1.00	24.85 hours
Moon and Sun	Earth	1.454	1.206	. . .
Earth	Moon	36.4	1325	stopped
Jupiter	Io	6760	4.5×10^7	stopped
Saturn	Titan	248	6.15×10^4	stopped
Neptune	Triton	706	4.99×10^5	stopped

[a] Not known; rotation period is perhaps of the order of 2 weeks.

rotation. From Table 9 it may be deduced that values of h^2 as large as 1.2 are not sufficiently high to stop a planet's rotation, while values of h^2 of approximately 2.0 or more appear to be large enough to stop a planet's rotation. It probably would not be too far afield to estimate a critical value of h^2 at approximately 2.0; that is, if h^2 is greater than 2.0, planetary rotation rates would probably be too slow (after the braking force had been working for several billion years) to be compatible with habitability. For a habitable planet to possess the proper surface temperatures in the vicinity of a small main-sequence star, however, it must orbit within an ecosphere—so close to the primary that the tidal braking effect becomes large. Thus for stars at the low-temperature end of the main sequence, the tidal braking effect and the planetary temperature requirements for habitability are incompatible.

The inner boundary due to tidal braking effect depends on both the mass of the primary and the properties of the planet, although it is quite insensitive to the planetary properties because $(R_B^4/M_B)^{\frac{1}{3}}$ is very nearly a constant for habitable planets. As may be seen in Figure 26, if h^2 equal to 2.0 is used as a criterion, habitable planets can exist in ecospheres only around stars having masses larger than 0.72 solar mass. A "full" ecosphere can exist around primaries of stellar mass greater than about 0.88 solar mass, but the ecosphere is narrowed by the tidal braking effect for primaries of lesser mass until it disappears when the stellar mass reaches about 0.72. The range in mass of stars that could have habitable planets is thus 0.72 to 1.43 solar masses, corresponding to main-sequence stars of spectral types F2 through K1. There is an extension of this range down to the larger class M stars (mass greater than 0.35 solar mass) for a

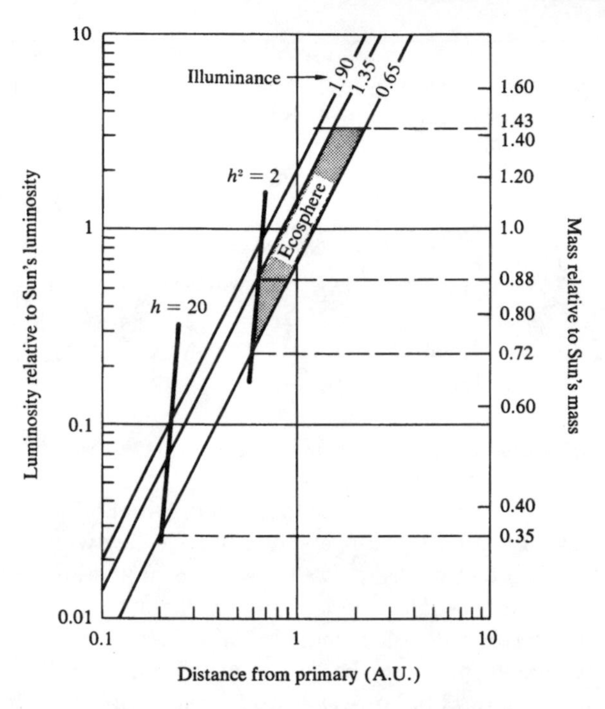

Figure 26. The boundaries of exospheres.

special class of planets with large satellites. This will be discussed in the next section.

SATELLITE RELATIONSHIPS

The tidal braking forces of a primary also apply to satellites. If h^2 of a satellite on a planet is greater than 2.0 but that of the primary is less than 2.0, one would expect to find the planet's rotation halted with respect to the satellite but continuing with respect to the primary. The planet's solar day and synodic month would be of the same length. For this condition to be compatible with habitability, however, the period would have to be such as to produce a solar day less than 96 hours in duration—a figure rather arbitrarily chosen as the longest day for habitability.

The relationship between a sidereal month P, the year Y, and the synodic month S, which may be written

$$\frac{1}{P} - \frac{1}{Y} = \frac{1}{S},$$

is such that to a rough approximation for small values of P relative to Y, S may be taken as equal to P. This simplifies the presentation of general conclusions.

If the tidal braking force due to a satellite, h_M^2, is greater than 2.0 and the relative rotation of the planet with respect to the satellite has stopped, the tides on the planet due to the satellite will be fixed (not moving across the surface of the planet), and thus they will not be an environmental variable. There may be tides due to the primary, however, and a new limiting condition will appear when the tides produced by the primary reach a destructive level incompatible with land life; that is, if the erosive power of tides becomes excessively high, all the dry land on the planetary surface will disappear, and the planetary surface will become a continuous deep ocean swept twice daily by tides of enormous magnitude. At what tidal magnitude would this occur? The Moon produces on the Earth mid-ocean tides approximately only a foot in height, yet local coastal tides are much higher because of the piling up of water in shallow bays; it might then be assumed that mid-ocean tides of the order of 10 to 20 feet would probably begin to be of sufficient magnitude to erode away all of the Earth's land masses over a period of many years. For present purposes, let us assume that the destructive tide limit is represented by h equal to 20.

Using this criterion, we can represent, by areas shown in Figure 27, all combinations of the tidal braking force due to the primary, h_S, and tidal braking force due to the satellite, h_M. In this figure, region 1 would contain freely rotating planets; region 2, planets with rotation halted with respect to a satellite; region 3, planets with rotation halted with respect to a satellite but with destructive tides due to the primary; and region 4, planets with rotation halted with respect to the primary. All habitable planets must fall within regions 1 and 2; and those falling within region 2 would be habitable only if their periods of rotation were less than 96 hours.

For a planet having the characteristics of the Earth, the limitations on satellite mass and distance are shown in Figure 28. The line marked "Roche's limit" marks a region inside which the smaller of the two bodies would tend to break into fragments as a result of the tide-raising forces of the larger.

It is interesting to note that within a certain range of satellite masses, twin habitable planets that did not rotate with respect to each other could exist.

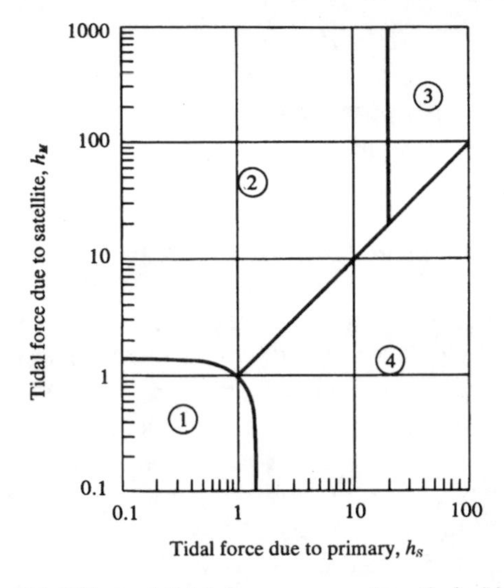

Figure 27. Effect of tidal forces on a planet's habitability.

The right-hand portion of Figure 28 represents the situation in which the satellite is larger than the habitable planet. For example, the vertical line marked "A" represents the case in which the satellite has ten times the mass of Jupiter and the habitable planet is revolving about it with its rotation halted. Somewhere in the vicinity of this line, a new boundary, determined by the heat produced by the more massive body, would be reached. As a satellite approaches stellar mass, its surface presumably becomes hotter because it has not been able to lose all of its heat of gravitational accretion and because it is also receiving heat from a primary. At some mass level a satellite would be hot enough to cause the loss of water from the atmosphere of the planet revolving about it, making the entire planet uninhabitable. Since we have little knowledge of the surface temperatures of bodies ten times as massive as Jupiter, this boundary at present can be located only approximately, as indicated by line A.

As we have seen, for stars at the low-temperature end of the main sequence, there is an incompatibility between the tidal braking effect and the temperature requirements of a habitable planet. The lower limit of mass of stars that could have freely rotating planets within an ecosphere was determined to be 0.72 solar mass. In this section we see a mechanism whereby the lower limit on stellar mass can be reduced still further. If a planet had a large, close satellite that maintained the planet's rotation rate so that the planet's solar day was shorter than 96 hours, it could

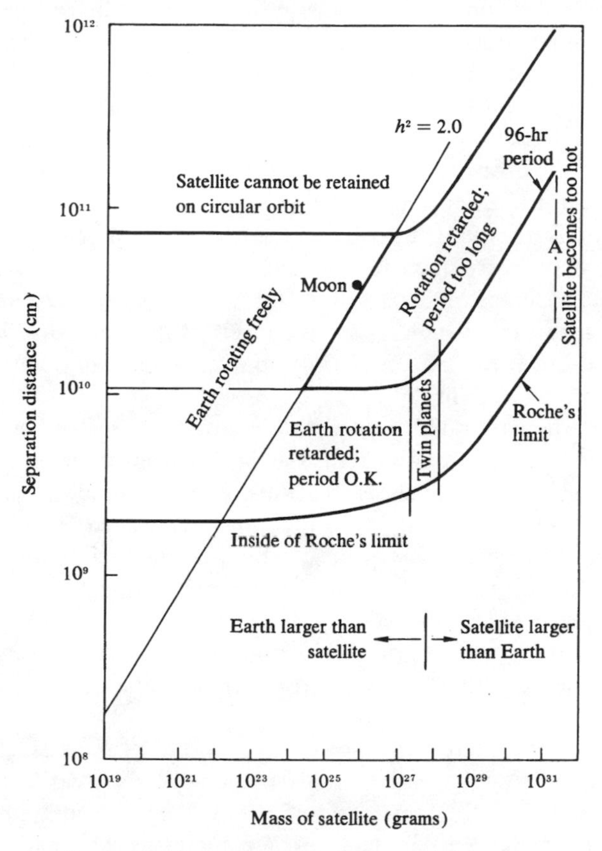

Figure 28. Hypothetical satellites of the Earth.

orbit within the ecosphere of stars less massive than 0.72 solar mass. However, a new limit would be reached when the tides on the planet due to the primary reached a destructive level. If we assume that the destructive tide level is 20 feet, then the new lower limit on stellar mass would be 0.35 solar mass. The dynamic stability of a system near the limit would probably be quite difficult to assess and the evolutionary sequence of events leading up to such a situation would be difficult to reconstruct.

SPECIAL PROPERTIES OF BINARY STAR SYSTEMS

So far, we have been concerned almost entirely with planets revolving about isolated stars. As will be discussed later, however, a large fraction of all stars exist in multiple stellar systems—binary, triple, and quadruple. The most common type of multiple system is the binary one; that is, where more than two stars are gravitationally associated, the other members are far removed from the first two—so far away that they can have little effect (from our point of view) on the binary system. Thus, if we examine the special properties of binary star systems important to habitable planets, in addition to the properties of isolated stars, then all the important classes of stars will have been covered.

Two essential questions must be analyzed with respect to the existence of habitable planets in binary star systems: (1) Can stable planetary orbits exist at the proper distances from a star in binary star systems? (2) If they can exist at such distances, are the levels of illumination (the sum of the illumination from two radiating sources) constant enough to be consistent with habitability?

Because of the vast number of possible combinations of astronomical parameters in a binary star system, the problem must be simplified for analysis. Some of the parameters that must be considered are the spectral types and masses of the two stars, their mean distance (semimajor axis of their orbit), the eccentricity of their orbit, and the inclination of planetary orbits with respect to the plane of the stellar orbit. At the outset, let us limit the problem to cases in which the two stars move in circles around their common center of gravity and the planet is in a near-circular orbit about one or both of the stars in the same plane as the two massive bodies.

This is now the restricted three-body problem of classical celestial dynamics, and regions in which stable planetary orbits can exist can be described in terms of the mass ratio of the two stars. For example, refer to Figure 29, stable near-circular direct planetary orbits can exist only inside boundaries A and D and outside boundary F. (For present purposes, retrograde planetary orbits can be ignored as representing less probable, and more limiting, circumstances.)

If at least one of the regions in which stable planetary orbits can exist also includes an ecosphere, then one requirement for the existence of habitable planets in a given binary star system is fulfilled.

The other requirement is that the level of illumination from the two radiating sources be fairly constant at the planetary orbit. How much variability would be permissible is a very difficult problem to assess because of the inadequacy of our knowledge concerning the prediction of planetary surface temperature from astronomical parameters and because the possible combinations of planetary variables are so numerous. In order to make an estimate, however, let us assume that the permissible variation in illumination for any given planet on a near-circular orbit must be less than 10 per cent during the annual period. Use of this figure

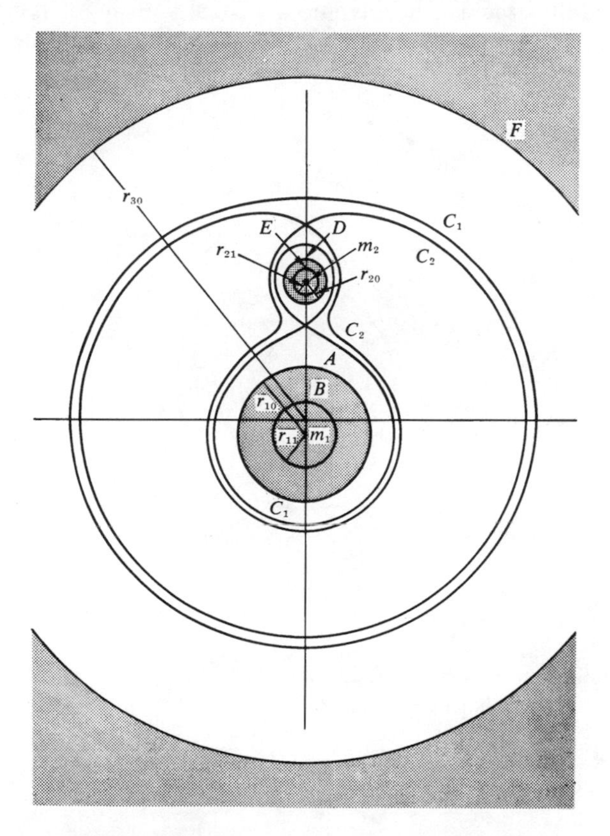

Figure 29. Regions within which stable, direct, near-circular planetary orbits can exist, $\mu = m_2/(m_1 + m_2) = 0.1$.

can not produce any great errors in the total picture of habitability of planets in the vicinity of binary stars.

Examples of illumination contours around two radiating sources are shown in Figure 30. As may be seen from this figure, there are regions in the immediate vicinity of each star where the illumination contours are nearly circular. Thus a planet in a near-circular orbit within these regions would receive a nearly constant supply of radiation. Also, at a great distance from both stars, the illumination contours are again nearly circular and are concentric to the larger source. But a planet in a near-circular orbit around both stars would move so that the center of its orbit was at the center of mass of the two stars; hence the planet would tend to cut across illumination contours unless it was quite far away. For the case illustrated, if we require that orbits must be stable and that the variability of illumination must not exceed 10 per cent, planetary

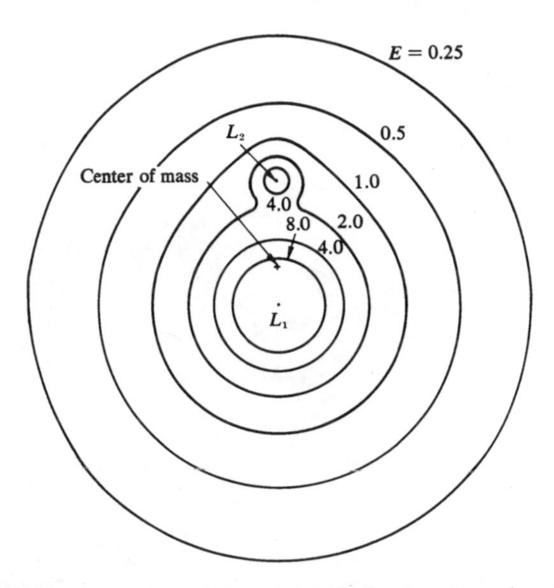

Figure 30. Contours of constant total illumination E in the vicinity of two radiating sources, $L_2/L_1 = 0.04$ and $\mu = 0.3$.

orbits meeting these requirements can exist within the following regions: around the more luminous star, inside a circle of radius 0.319 unit, as limited by stability; around the less luminous star, inside a circle of radius 0.108 unit, as limited by variability of illumination limits; around both stars, outside a circle of radius 10.5 units, as determined by variability of illumination limits (see Table 10 and Figure 31).

For habitability, however, the absolute mean levels of illumination must fall within certain limits, as discussed previously. For a given value of L_1 (the luminosity of the more luminous star) and λ (the relative luminosity of the less luminous star), it is possible to compute the minimum separation distances for binary star systems in order to have habitable planets about either star and the maximum separation distance in order to have habitable planets in orbit about both stars together.

***Table 10. Boundaries of Regions in the Vicinity of Two Luminous Sources Where Habitable Planets Can Exist in Near-circular Orbits*[a]**

Binary characteristics		Planets of larger source		Planets of smaller source		Planets in orbit about both sources	
luminosity ratio, L_2/L_1	assumed mass ratio, μ	dynamical limit, r_{10}	illumination limit, r_{13}	dynamical limit, r_{20}	illumination limit, r_{23}	dynamical limit, r_{30}	illumination limit, r_{33}
1	0.5	0.257	0.286	0.257	0.286	2.165	3.16
0.5	0.455	0.270	(*b*)	0.245	0.233	2.19	5.02
0.1	0.345	0.300	(*b*)	0.220	0.143	2.31	10.12
0.04	0.301	0.319	(*b*)	0.206	0.1084	2.41	10.48
0.01	0.243	0.335	(*b*)	0.188	0.0724	2.41	9.31
0.001	0.1411	0.395	(*b*)	0.157	0.0403	2.31	5.60
10^{-4}	0.0826	0.445	(*b*)	0.135	0.0282	2.17	3.30
10^{-5}	0.0468	0.500	(*b*)	0.113	0.0254	1.98	1.87
10^{-6}	0.0262	0.560	(*b*)	0.0930	0.0250	1.80	1.04

[a] Limiting radius is underlined.
Maximum orbital radius for stable orbit: r_{10}, r_{20}. Minimum orbital radius for stable orbit: r_{30}. Beyond limits r_{13} and r_{23}, illumination varies by more than 10 per cent. Inside limit r_{33}, illumination varies by more than 10 per cent.
For the purposes of this table, it is assumed that L is proportional to $M^{3.8}$.
[b] The radius r_{13} is greater than the limit r_{10}.

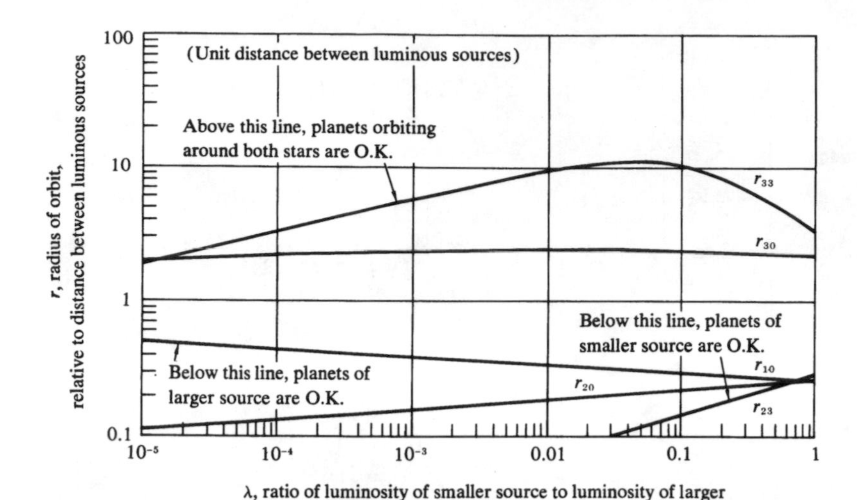

Figure 31. Regions in the vicinity of two radiating sources where planets can exist on stable, near-circular orbits and where total illumination varies less than 10 per cent.

SUMMARY OF THE BASIC REQUISITES OF A HABITABLE PLANET

Its *mass* must be greater than 0.4 Earth mass, to permit it to produce and retain a breathable atmosphere, and less than 2.35 Earth masses, since surface gravity must be less than 1.5 *g*.

Its *period of rotation* must be less than about 96 hours (4 Earth days) to prevent excessively high daytime temperatures and excessively low nighttime temperatures.

The *age* of the planet (and the star about which it orbits) must be greater than about 3 billion years, to allow for the appearance of complex life forms and the production of a breathable atmosphere.

Its *axial inclination* (inclination of its equator to the plane of its orbit) and the *level of illumination* from its sun are interrelated and determine the temperature patterns on its surface. Generally speaking, levels of illumination at low inclinations should lie between 0.65 and 1.35 times Earth normal (roughly 10 to 20 lumens per square centimeter), although certain combinations of illumination up to 1.9 times Earth normal and inclinations up to 81 degrees are compatible with marginal habitability.

The *orbital eccentricity* must be less than approximately 0.2, since greater values of eccentricity would produce unacceptably extreme temperature patterns on the planetary surface.

The *mass of the primary* must be less than 1.43 solar masses, since its residence time on the main sequence must be greater than 3 billion years. The mass of the primary must be greater than 0.72 solar mass because, when primaries have masses smaller than this, there is an incompatibility between permissible illumination levels and tidal retardation of a planet's rotation. For a special, rare class of planets with extremely large or close satellites, there is an extension of the lower permissible primary mass down to 0.35 solar mass.

If the planet orbits in a *binary star system*, the two stars must either be quite close together or quite far apart so as not to interfere with the stability of the planetary orbit and not to produce too variable a level of illumination at the planetary distance.

If all of these requisites are satisfied, then there is a very good possibility that a planet will be habitable.

CHAPTER 5

Probability of Occurrence of Habitable Planets

Having summarized the properties of habitable planets and the astronomical requirements implied by these properties, we can now attempt to estimate the prevalence of such bodies in our Galaxy (the Milky Way); and to do this with any reasonable degree of accuracy (in the spirit of the present study), it is necessary to consider the following factors:

N_s, the prevalence of stars in the suitable mass range, 0.35 to 1.43 solar masses;

P_p, the probability that a given star has planets in orbit about it;

P_i, the probability that the inclination of the planet's equator is correct for its orbital distance;

P_D, the probability that at least one planet orbits within an ecosphere;

P_M, the probability that the planet has a suitable mass, 0.4 to 2.35 Earth masses;

P_e, the probability that the planet's orbital eccentricity is sufficiently low;

P_B, the probability that the presence of a second star has not rendered the planet uninhabitable;

P_R, the probability that the planet's rate of rotation is neither too fast nor too slow;

P_A, the probability that the planet is of the proper age;

P_L, the probability that, all astronomical conditions being proper, life has developed on the planet.

Once values for all of these factors have been established, the estimated number of habitable planets N_{HP} in the Galaxy can be expressed as the product:

$$N_{HP} = N_s P_p P_i P_D P_M P_e P_B P_R P_A P_L.$$

Obviously any such number is bound to be highly imprecise, since not all of the factors in the above equation are known with accuracy. In fact, most of them can only be estimated rather roughly and the values assigned to some of them depend on which cosmogonical theory one happens to favor. Therefore, the reader should not infer that the estimates contained in this chapter and the next are intended to represent the final word on the subject. They merely represent one person's attempt to arrive at numbers having some rationale behind them. The reader is welcome to alter them in any way that better conforms to his own view of the universe.

PREVALENCE OF STARS IN SUITABLE MASS RANGE, 0.35 TO 1.43 SOLAR MASSES

Counts of the numbers of stars of the various magnitudes and spectral types in the neighborhood of the Sun have been made by astronomers. From these counts and from the relationships between magnitude and mass and between mass and spectral type, it is possible to estimate the concentration of stars of each spectral type in the solar neighborhood. These data are presented in Figures 32 and 33 and in Table 11.

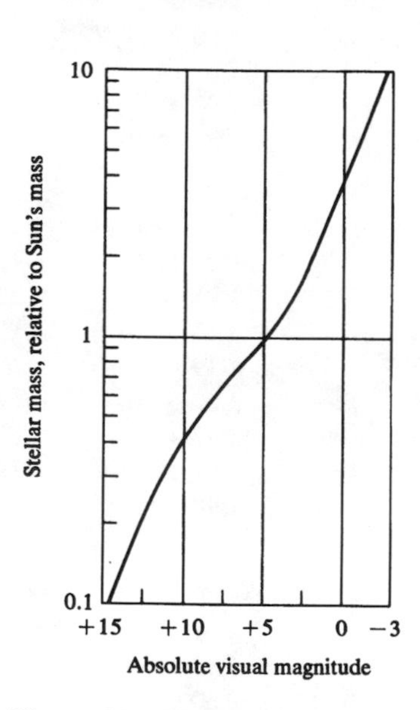

Figure 32. Absolute visual magnitude of main-sequence stars as a function of stellar mass.

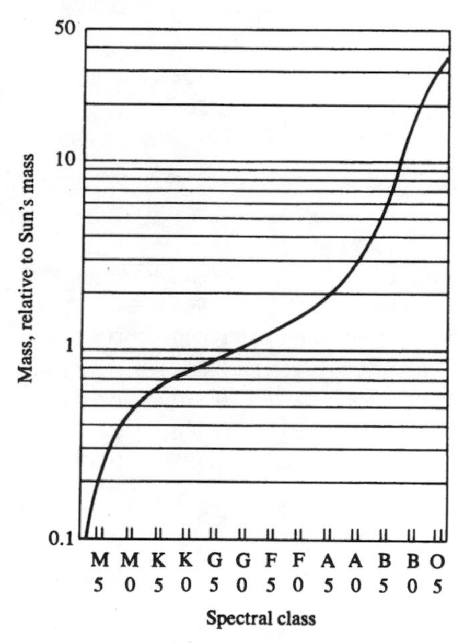

Figure 33. Mass as a function of spectral class (main-sequence stars).

Table 11. Characteristics

Spectral class	Mass range[a]	Mean mass,[a] M	Luminosity,[a] L	Absolute visual magnitude	Radius,[a] R	Number per cubic parsec[b]	Residence on main sequence, τ_{ms} (10^9 years)
F0	1.51 –1.57	1.54	4.46	+2.88	1.40	2.05×10^{-4}	2.12
1	1.44 –1.51	1.47	3.85	3.05	1.35	2.48×10^{-4}	2.66
2	1.37 –1.44	1.40	3.02	3.26	1.30	2.76×10^{-4}	3.47
3	1.32 –1.37	1.34	2.82	3.40	1.26	3.20×10^{-4}	3.98
4	1.26 –1.32	1.28	2.32	3.61	1.22	3.66×10^{-4}	4.90
5	1.21 –1.26	1.24	2.07	3.77	1.19	3.92×10^{-4}	5.37
6	1.16 –1.21	1.18	1.72	4.01	1.14	4.04×10^{-4}	6.46
7	1.12 –1.16	1.14	1.55	4.10	1.10	4.12×10^{-4}	7.25
8	1.07 –1.12	1.10	1.37	4.29	1.08	4.25×10^{-4}	8.32
9	1.04 –1.07	1.06	1.22	4.50	1.05	4.36×10^{-4}	9.34
G0	1.00 –1.04	1.02	1.04	4.60	1.01	4.43×10^{-4}	10.2
1	0.970–1.00	0.985	0.956	4.80	0.99	4.52×10^{-4}	11.9
2	0.945–0.970	0.955	0.814	4.96	0.97	4.55×10^{-4}	>12
3	0.918–0.945	0.930	0.694	5.15	0.95	4.99×10^{-4}	>12
4	0.888–0.918	0.900	0.610	5.30	0.92	5.35×10^{-4}	>12
5	0.860–0.888	0.870	0.525	5.53	0.90	6.40×10^{-4}	>12
6	0.839–0.860	0.850	0.501	5.65	0.88	6.90×10^{-4}	>12
7	0.810–0.839	0.825	0.408	5.85	0.86	7.60×10^{-4}	>12
8	0.783–0.810	0.800	0.363	5.90	0.84	7.90×10^{-4}	>12
9	0.762–0.783	0.775	0.316	6.12	0.82	8.35×10^{-4}	>12
K0	0.740–0.762	0.750	0.282	6.40	0.80	8.74×10^{-4}	>12
1	0.719–0.740	0.730	0.252	6.60	0.78	9.14×10^{-4}	>12
2	0.692–0.719	0.705	0.219	6.95	0.76	9.72×10^{-4}	>12
3	0.670–0.692	0.680	0.200	7.20	0.74	1.01×10^{-3}	>12
4	0.640–0.670	0.655	0.162	7.50	0.72	1.11×10^{-3}	>12
5	0.615–0.640	0.630	0.145	7.90	0.70	1.19×10^{-3}	>12
6	0.585–0.615	0.600	0.123	8.00	0.68	1.30×10^{-3}	>12
7	0.560–0.585	0.570	0.105	8.50	0.652	1.47×10^{-3}	>12
8	0.526–0.560	0.540	0.0912	8.75	0.635	1.65×10^{-3}	>12
9	0.490–0.526	0.505	0.0726	9.10	0.608	1.98×10^{-3}	>12
M0	0.452–0.490	0.470	0.0596	9.48	0.585	2.42×10^{-3}	>12
1	0.408–0.452	0.435	0.0486	9.80	0.560	3.00×10^{-3}	>12
2	0.353–0.408	0.380	0.0317	10.40	0.525	4.31×10^{-3}	>12
3	0.300–0.353	0.330	0.0232	10.9	0.490	6.39×10^{-3}	>12
4	0.247–0.300	0.270	0.0152	11.7	0.455	9.50×10^{-3}	>12
5	0.200–0.247	0.220	0.0120	12.3	0.420	1.23×10^{-2}	>12

[a] Relative to Sun.

[b] In our spiral Galaxy (the Milky Way); average stellar densities. Densities are higher near the

of Main-sequence Stars

		Ecosphere bounds						
Effective surface temperature, T (°K)	Density, ρ (g/cm³)	inner, D_1 $E=1.35$, $h^2 \leq 2.0$ (A.U.)	outer, D_2 $E=0.65$ (A.U.)	P_A	P_D	P_{HP} isolated	P_{HP} general	Number of stars with habitable planets per cubic parsec[b]
7060	0.791	1.82	2.62	0	0.63	0	0	0
6930	0.842	1.69	2.43	0	0.63	0	0	0
6650	0.898	1.50	2.16	0.1355	0.63	0.0111	0.0106	2.93×10^{-6}
6640	0.945	1.44	2.08	0.246	0.63	0.0202	0.0192	6.15×10^{-6}
6425	0.994	1.31	1.89	0.388	0.63	0.0318	0.0303	1.11×10^{-5}
6320	1.04	1.24	1.78	0.441	0.63	0.0362	0.0344	1.35×10^{-5}
6170	1.12	1.13	1.63	0.536	0.63	0.0440	0.0418	1.69×10^{-5}
6120	1.21	1.07	1.54	0.586	0.63	0.0481	0.0457	1.88×10^{-5}
5990	1.23	1.01	1.45	0.640	0.63	0.0525	0.0499	2.12×10^{-5}
5900	1.29	0.951	1.37	0.679	0.63	0.0556	0.0529	2.31×10^{-5}
5780	1.40	0.878	1.26	0.700	0.63	0.0574	0.0545	2.41×10^{-5}
5710	1.43	0.842	1.21	0.700	0.63	0.0574	0.0545	2.46×10^{-5}
5545	1.48	0.776	1.12	0.700	0.63	0.0574	0.0545	2.48×10^{-5}
5380	1.53	0.717	1.03	0.700	0.63	0.0574	0.0545	2.72×10^{-5}
5300	1.63	0.675	0.970	0.700	0.63	0.0574	0.0545	2.91×10^{-5}
5160	1.68	0.640	0.900	0.700	0.60	0.0546	0.0520	3.33×10^{-5}
5160	1.76	0.630	0.880	0.700	0.59	0.0538	0.0511	3.53×10^{-5}
4955	1.83	0.626	0.795	0.700	0.455	0.0415	0.0394	2.99×10^{-5}
4870	1.90	0.618	0.746	0.700	0.36	0.0328	0.0312	2.46×10^{-5}
4760	1.98	0.610	0.700	0.700	0.255	0.0232	0.0221	1.84×10^{-5}
4685	2.06	0.604	0.662	0.700	0.17	0.0155	0.0147	1.28×10^{-5}
4610	2.17	0.599	0.625	0.700	0.07	0.00638	0.00606	5.54×10^{-6}
4560	2.26	...	...	0.700	0	0	0	0
4470	2.37	...	...	0.700	0	0	0	0
4300	2.48	...	...	0.700	0	0	0	0
4240	2.50	...	...	0.700	0	0	0	0
4130	2.69	...	...	0.700	0	0	0	0
4050	2.90	...	...	0.700	0	0	0	0
3965	2.97	...	...	0.700	0	0	0	0
3830	3.17	...	...	0.700	0	0	0	0
3720	3.31	...	...	0.700	0	0	0	0
3600	3.49	...	...	0.700	0	0	0	0
3350	3.70	...	...	0.700	0	0	0	0
3210	3.96	...	...	0.700	0	0	0	0
2990	4.04	...	...	0.700	0	0	0	0
2935	4.19	...	...	0.700	0	0	0	0
								Total 4.033×10^{-4}

galactic center and in globular clusters; they are lower near the galactic rim.

From Table 11, it may be seen that the suitable mass range, 0.35 to 1.43 solar masses, embraces all main-sequence stars in the spectral class range from approximately M2 to F2. The estimated numbers of stars in each spectral class (per cubic parsec* of space) are given. It is clear that the prevalence of stars in a given mass range is a strong function of mass, the less massive stars being the most abundant and the very massive stars being quite rare. This is brought out more strikingly in Table 12.

Table 12. Relative Abundances of the Classes of Main-sequence Stars

Spectral class	Approximate mass relation to Sun	Main sequence (per cent)[a]
O	32	0.0000186
B	6	0.0929
A	2	0.581
F	1.25	2.905
G	0.9	7.315
K	0.6	15.09
M	0.22	73.25
		Total 99.23

[a] This column represents the percentage of all stars. Stars not on the main sequence (primarily giants) make up the remaining 0.77 per cent. For these stars, spectral class and mass are not directly related.

The giants, or stars that have left the main sequence, are seen to account for less than 1 per cent of all stars, although they are extremely prominent in the night sky because of their great luminosity. Those stars of major interest to us as possible primaries in a system containing habitable planets are relatively inconspicuous objects having absolute visual magnitudes in the range from about +7 to +3. Absolute magnitude is a measure of the appearance a star would have if viewed at a distance of 10 parsecs, or 32.63 light-years. The absolute magnitude of our Sun, for example, is +4.8, while the faintest stars normally visible to the naked eye in the night sky are of +5 or +6 apparent magnitude. On the other hand, of the one hundred brightest stars in the sky, all but one have absolute magnitudes in the range +3 to −7. The exception is Alpha Centauri

* One parsec equals 3.26 light-years; one cubic parsec equals 34.7 cubic light-years.

(absolute visual magnitude, +4.5), which appears bright to us only because of its nearness. From Table 12 it may be seen that the most important stars (in the present context), those main-sequence stars in spectral classes F, G, and K, account for about 25 per cent of all stars. They are not at all rare objects.

Since the mass of a star also determines its lifetime in the main sequence and has a bearing on the size of its ecosphere, the prevalence of stars in the proper mass range N_s must be considered in conjunction with several other parameters to determine the number of habitable planets in the Galaxy.

PROBABILITY THAT A GIVEN STAR HAS PLANETS IN ORBIT AROUND IT

On the basis of ideas currently held about the formation of stars within clouds of dust and gas, it is reasonable to postulate here that *every* star, at least every star in the mass range of interest, has a family of planetary bodies in orbit about it. Some small fraction of the total retained mass of the original cloud should always have a velocity relative to the central star sufficiently high to prevent its being captured by the star. The magnitude of this fraction probably depends on thc original angular velocity of the cloud and on its mass. It is believed by some authorities that the initial angular speed of a diffuse cloud depends in turn on its distance from the center of the Galaxy. Since it is extremely improbable that the original angular velocity of the cloud would be precisely zero, it is highly probable that some material would be left in orbit after the central star had completed its growth. For present purposes P_p will be taken as 1.0.

Some earlier ideas about the formation of the solar system (which are still current in some quarters) depend on a catastrophic origin—a near collision between the Sun and a passing star. This hypothesis requires the planets to have been formed from incandescent material pulled out of one or both of the two stars as they passed each other. From a dynamic standpoint, it is quite improbable that any material would remain in orbit after such an encounter; the material drawn out in an incandescent tidal filament between the stars would tend to fall back into the stars once they had separated. Moreover, even if such incandescent material did remain in orbit, it would have no tendency to coalesce into the form of large droplets, as has often been hypothesized; instead it would literally explode into its constituent atoms, since the velocity of the hydrogen atoms, the main constituent, at stellar temperatures would be many times greater than the velocity of escape from the hypothetical filament.

The result would be a sun surrounded by a flat ring of gas and dust. (Within this ring, planets might subsequently form by accretion, but this is not the usual chain of events hypothesized in the catastrophic mode of formation.) One consequence of the tidal filament, near-collision mode of planetary system formation is that such systems would be rare in the Galaxy. The theory of noncatastrophic, or normal, mode of formation implies that planetary systems are the rule, rather than the exception.

PROBABILITY THAT THE INCLINATION OF THE PLANET'S EQUATOR IS CORRECT FOR ITS ORBITAL DISTANCE

As we have seen, the surface-temperature pattern of a planet depends mainly on the intensity of the illumination received at the mean distance

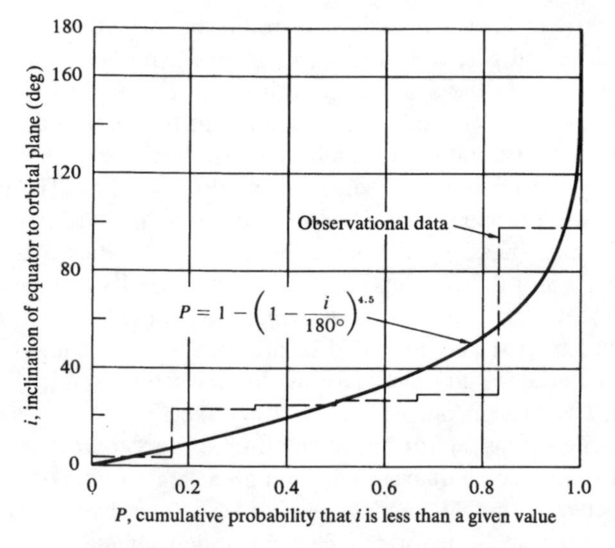

Figure 34. Adopted curve for estimating the probability that the inclination of a planet's equator is less than a given value.

from its primary, and on the inclination of its equator. A fairly complicated relationship exists between habitability, mass of primary, mass of planet, and mean distance from primary with respect to the retardation of rotation due to tidal effects at the inner edge of the ecosphere. When associated with small primaries, large planets have wider ecospheres than small planets, since, being larger, they have higher rotational energies and are not so rapidly retarded.

Because of the complexity of the relationship mentioned above, it would be complicated to an unwarranted degree to compute the probability of planet habitability for all possible combinations of inclination and illumination, together with all possible combinations of mass of primary, planetary mass, and distance. In the spirit of the present study, where the emphasis is not so much on precision as on obtaining certain rough but reasonable estimates, some simplification is desirable. Therefore it was decided, for purposes of calculation, to set separate limits on allowable illumination range and on allowable inclination.

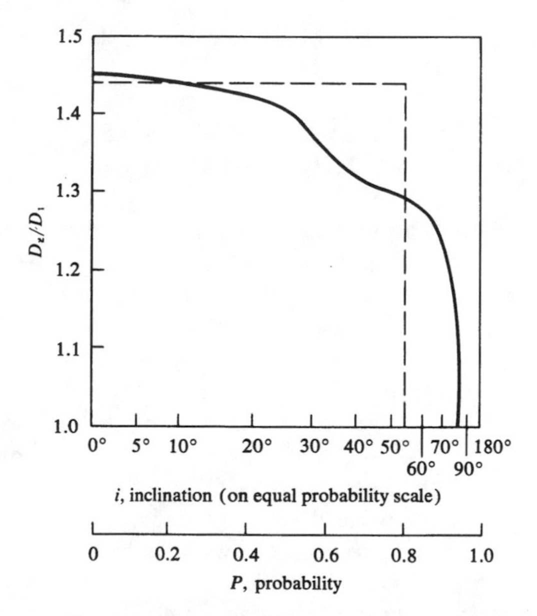

Figure 35. Relative width of ecosphere as a function of inclination of planet's equator.

Using data on the equatorial inclinations of the planets of the solar system, an assumed cumulative probability curve was chosen of the form, $P = 1 - (1 - i/180°)^n$, with P being the probability that a randomly selected planet will have an equatorial inclination less than a given value i. A satisfactorily good fit was obtained for $n = 4.5$, as shown in Figure 34.

This general relationship was applied to the habitability regions on the illumination-versus-inclination diagram (see Figure 24 on page 66) to produce the probability-of-occurrence pattern of Figure 35. For simplification, the curve-bounded region of Figure 24 was replaced by a rectangle having the same area. For illumination limits of E_1 (level at inner edge of ecosphere) and E_2 (level at outer edge), related to each other so that

$$\sqrt{\frac{E_1}{E_2}} = \frac{D_2}{D_1} = 1.441,$$

the required value of P is about 0.81. This corresponds to an inclination of about 54 degrees as an upper limit.

Choosing $P_i = 0.81$ over the whole range of tolerable illumination levels as an approximation, rather than letting P_i be a function of illumination level, greatly simplifies the subsequent calculations without introducing any great errors. Thus, it will be assumed that $P_i = 0.81$.

PROBABILITY THAT AT LEAST ONE PLANET ORBITS WITHIN THE ECOSPHERE OF A GIVEN STAR

As discussed earlier, primaries having masses greater than 0.88 solar mass have complete ecospheres, while those with masses between 0.72 and 0.88 solar mass have narrowed ecospheres because of the rotation-retarding effects of the primary at the inner edge of the ecosphere. To estimate the probability that there will be at least one planet orbiting within an ecosphere, it is necessary to know something about the spacing of planets in planetary systems. Unfortunately there is only one planetary system available to us for study, our own solar system. Since the planets in our system are spaced in a manner that conforms with criteria for long-term orbital stability, however, it seems justifiable to assume that other planets are spaced in accordance with the same laws that have operated here. The spacing of the orbits of planets in the solar system may be seen to be compatible with permitted nearly circular orbits in the restricted three-body problem. The only interference is between the orbits of Neptune and Pluto and this is not strictly an interference because of the inclination of Pluto's orbit. The remainder of the solar system is approximately half taken up with planetary orbits in such a manner that there are comfortable gaps between them. It is assumed that this pattern may well be a universal feature of planetary systems around other stars.

Using this assumption based on our solar system, it is now possible to compute the probability that at least one planet orbits in the distance interval D_i to D_o. Taking the actual mean orbital distances from the Sun of the planets of the solar system (see Figure 18 on page 51), the question was asked, "What is the probability that a given distance interval, D_i to D_o, placed at random, will encompass at least one planetary orbit?" For example, for the ratio D_o/D_i equal to 3.41 or greater in the solar system, the interval D_i to D_o can not be placed so that it does not include at least one planet, since the widest gap between planets on a logarithmic scale is between Jupiter at 5.20 astronomical units and Mars at 1.524 astronomical units and the ratio of these distances is 3.41. Hence, for D_o/D_i equal to 3.41, the probability that one or more planets will exist in the interval D_i to D_o is 1.0. For lesser values of D_o/D_i, the probabilities are less than 1.0. Complete results for the solar system are given in Figure 36.

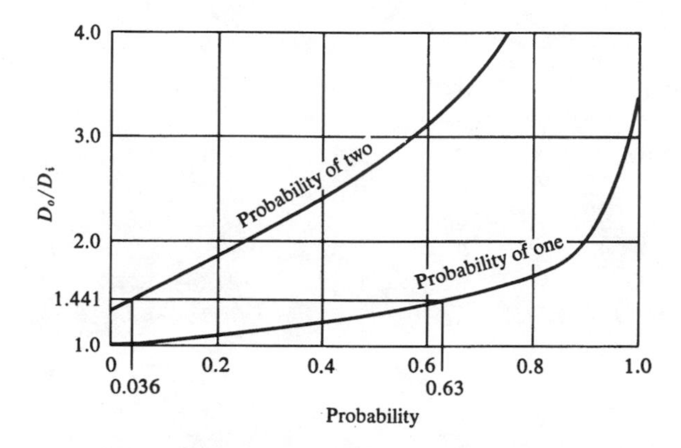

Figure 36. The probability that at least one planet exists in the distance interval D_i to D_o (based on solar-system data).

If we take $D_o/D_i = 1.441$ $[=(1.35/0.65)^{\frac{1}{2}}]$ for a complete ecosphere between the illumination limits 0.65 and 1.35 solar constants, then it may be seen that the probability of at least one planet orbiting within a complete ecosphere is 0.63. For smaller values of D_o/D_i, as for primaries with narrowed ecospheres in the mass range 0.72 to 0.88, the probability of the existence of one planet is correspondingly reduced. For each type of primary in the mass range of interest, the probability of the existence of one planet is given in Table 13.

PROBABILITY THAT THE PLANET HAS A SUITABLE MASS

The number-versus-mass distribution for the planets of the solar system also exhibits a pattern of regularity. The smaller individuals among the planets are more numerous than the larger ones, as is the case with stars, asteroids, and meteorites. This is apparent from Figure 10 (see page 33). Under the assumption that the planets of our solar system represent a fair sampling of such objects, it is now possible to construct a probability

Table 13. Probability That a Planet Orbits within an Ecosphere

Spectral class	Mean mass	Luminosity	Ecosphere bounds inner, D_1	Ecosphere bounds outer, D_2	D_2/D_1	P_D
F0	1.54	4.46	1.82	2.62	1.441	0.63
G0	1.02	1.04	0.878	1.26		
G4	0.900	0.610	0.675	0.970	1.441	0.63
5	0.870	0.525	0.640	0.900	1.407	0.60
6	0.850	0.501	0.630	0.880	1.397	0.59
7	0.825	0.408	0.626	0.795	1.270	0.455
8	0.800	0.363	0.618	0.746	1.209	0.36
9	0.775	0.316	0.610	0.700	1.147	0.255
K0	0.750	0.282	0.604	0.662	1.097	0.17
1	0.730	0.252	0.599	0.625	1.043	0.07
2	0.705	0.219	...	...	...	0

distribution for the planets of our solar system. Such a distribution was made by smoothing the data presented in the curve of Figure 10 and dividing the mass range from 0.03 to 3000 Earth masses into small mass increments, each increment having associated with it a number such that the sum of the numbers over the entire mass range equalled nine (the number of planets in our solar system). The probability that a planet chosen at random lies within any given mass interval is one-ninth the sum of the numbers associated with the mass increments making up that mass interval. Some of these probabilities are shown in Figure 37. From this curve, the probability that a planet chosen at random lies within the mass range 0.4 to 2.35 Earth masses is seen to equal 0.19. Thus, for present purposes, P_M will be taken as 0.19.

From some points of view, this may appear to be an unnecessarily pessimistic approach since, in our system, there is a rough, though by no means regular, distribution of planetary mass as a function of distance from the Sun. The small terrestrial planets are close in; the massive giant planets are far out. This may suggest a general law that could be applied to all systems; but, because there are so many irregularities in the pattern and, since attempts to explain the mass distribution have not been very successful, it was thought preferable here to adopt the idea that planetary mass is independent of distance from the primary. Our concepts of the

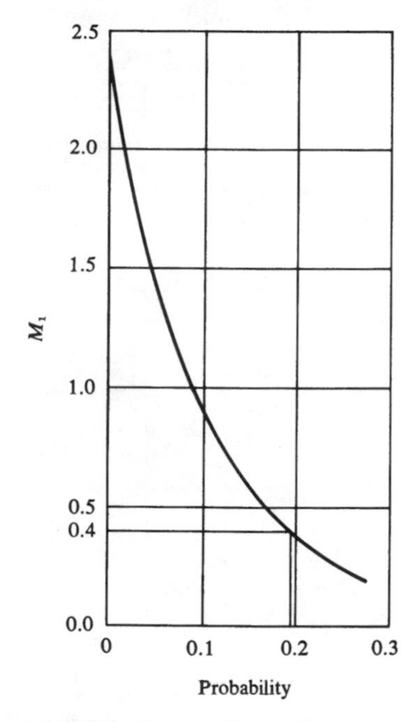

Figure 37. The probability that a planet chosen at random lies within the mass range from M_1 to 2.35 times the mass of the Earth.

processes of solar and planetary formation and development are, after all, quite incomplete at this time.

PROBABILITY THAT THE PLANET'S ORBITAL ECCENTRICITY IS SUFFICIENTLY LOW

An inspection of the values of orbital eccentricity among the bodies of planetary size in the solar system suggests that these eccentricities are distributed according to an empirical probability curve having the approximate form $P = 1 - (1 - e)^N$, where N is equal to about 13. In this assumed distribution, P is the cumulative probability that the eccentricity is less than any assigned value. Undoubtedly, other empirical distribution curves could be found that would fit the observational data more closely than the distribution given here. It was not considered worth while, however, to develop a better fit because of the sparseness of data.

If, as already discussed, we permit orbital eccentricities up to 0.2, then, as shown in Figure 38, about 94 per cent of planets should have eccentricities below this value and thus should be habitable, provided that all

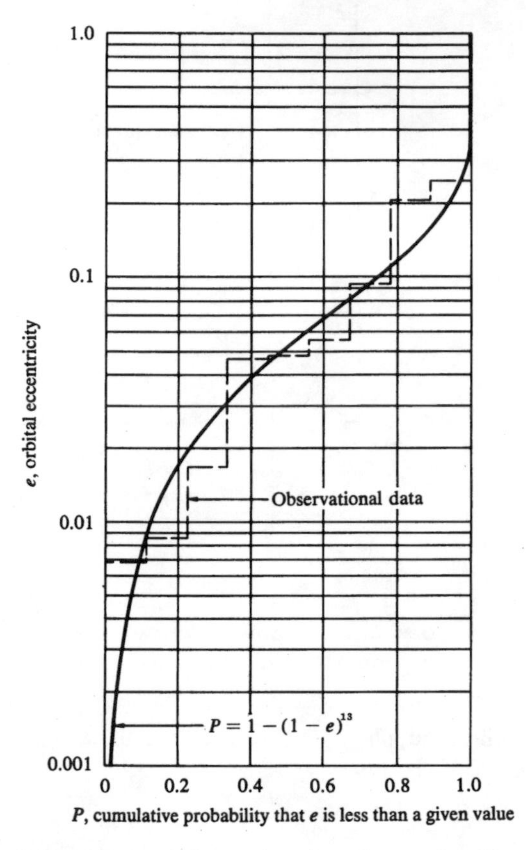

Figure 38. Adopted curve for estimating the probability that the eccentricity of a planet's orbit is less than a given value.

the other essential requirements are met. For present purposes, we will take P_e equal to 0.94.

PROBABILITY THAT THE PRESENCE OF A SECOND STAR HAS NOT RENDERED THE PLANET UNINHABITABLE

In order to assess this probability, it is necessary to know the probability that a star is a member of a binary star system, and if so, the probability that the stars are spaced, or move around each other, in such a manner that they will affect the existence of a normal ecosphere.

As discussed earlier, habitable planets can exist in binary star systems if the two stars are so close together that there is a single ecosphere around the pair or if they are so far apart that at least one can have an ecosphere without interference from the other.

Tables 14 and 15 give the proportion of stars that are spectroscopic doubles (closely spaced binaries) as a function of spectral class. In spectral classes F, G, and K, the proportion is 4 to 7 per cent according to Allen (1955) and 28 to 30 per cent according to Jaschek and Jaschek (1957). The periods of revolution of spectroscopic doubles made up of main-sequence stars are typically measured in days. This means that the two stars must be quite close together, generally having separations of less than 0.1 astronomical unit. From Figure 31 (see page 80), it may be seen that this would be compatible with the existence of an ecosphere around both stars.

On the other hand, the great majority of visual doubles are so widely separated that their orbital elements can not be determined with much accuracy. For the 12 per cent of visual doubles with measurable orbital motions, typical separations are great enough to permit the existence of a suitable ecosphere around either star. Since 4 to 9 per cent of stars in spectral classes F, G, and K are visual doubles, probably no more than a fraction of 1 per cent have separations in the critical distance range that would prevent the existence of an ecosphere.

According to van den Bos (1956), the median period of the binaries near the Sun has been given by Hertzsprung as 80 years and by Luyten as 300 years. Wide separations appear to be typical.

Some authors have assumed that binary systems, unless the components are very widely separated, can not have stable planetary systems (Huang, 1959). Considerations of the stability limits for orbits in the three-body problem, however, strongly suggest that stable planetary orbits can exist within ecospheres in the majority of binary systems of interest, although the general conditions for stability of planetary orbits are difficult to estimate when the two stellar components are revolving in eccentric orbits about their common center of mass.

From the foregoing, it appears that practically all of the closely spaced binary star systems (spectroscopic doubles) could possess habitable planets and that, among visual doubles, spacings that would prevent the existence of a normal ecosphere are rare. In view of the incomplete state of our knowledge, however, let us estimate that generally, for any arbitrary star, there will be interference in 5 per cent of all cases, or P_B is equal to 0.95, to allow for possible underestimates of the numbers of binary star systems

Table 14. Data on Binary Stars

Spectroscopic doubles							
Spectral class	O	B	A	F	G	K	M
Proportion of the stars in each class with duplicity in radial velocity (per cent)	...	24	13	7	6	4	6
Median period of revolution (days) for spectroscopic doubles composed of main-sequence stars	4	4	5	6	10	10	1 (?)

Visual doubles (of apparent visual magnitude brighter than +9)							
Spectral class	O	B	A	F	G	K	M
Proportion of the stars in each class that are visual doubles (per cent)	...	2	8	6	9	4	2

Visual doubles with measurable orbital motions

Proportion of visual doubles with measurable orbital motions is 12 per cent (others are too widely separated for determination of orbital elements).

Mean and modal elements (*from* Arend, 1950)	mean	mode
Semimajor axis of true orbit (A.U.)	26	18
Period (years)	76	32
Eccentricity	0.54	0.46

Source: Allen (1955).

Table 15. Corrected Percentages of Spectroscopic Binaries along the Main Sequence

Spectral class	B	A	F	G	K	M
Mean mass	5.00	2.00	1.25	0.90	0.63	0.40
Percentage of spectroscopic binaries in each class	21	27	30	29	28	32

Source: Jaschek and Jaschek (1957).

having awkward separation distances. For isolated stars or for stars in binary systems where the orbital parameters are known to be favorable, P_B should be taken as 1.00.

PROBABILITY THAT THE PLANET'S RATE OF ROTATION IS NEITHER TOO HIGH NOR TOO LOW

As we have seen in Chapter 3, there seems to be a direct empirical relationship between planetary mass and rotational energy per unit mass for the planets of our solar system having unrestricted rotation. Similar relationships may well be found in other planetary systems, since the same evolutionary and dynamic forces would be working in a similar manner. Based on this assumption, planets in the mass range from 0.4 to 2.35 Earth masses that orbit within ecospheres would be expected to have rotational periods well within the desired range. Because of the sparseness of the data, however, and because we do not know with much confidence the factors that determine a planet's rotation rate, let us take P_R equal to 0.9 for tidally unretarded planets. Habitable close-in planets with large or close satellites will be considered too scarce to affect our calculations.

PROBABILITY THAT THE PLANET IS OF THE PROPER AGE

It was concluded previously that it takes a period of the order of 3 billion years to produce a habitable planet, provided all the astronomical conditions are correct. We would here like to estimate the probability that a given star (with its planets) is older than 3 billion years.

Since we know (or believe that we know approximately) the relationship between stellar mass and time of residence (total life expectancy) in the main sequence, τ_{ms}, we then know that a given star on the main sequence can not be older than τ_{ms}. It also probably can not be older than the galaxy in which it exists. We know too that stars are still being formed in our Galaxy. There is not at present, however, a reliable way of determining whether a given main-sequence star was formed fairly recently or whether it is nearing the end of its residence in the main sequence.

If we assume that stars are forming and have been formed during the past at a fairly constant rate, an expression can be obtained for the probability that a given star is older than, say, 3 billion years:

$$P_A = \frac{t-3}{t} \quad \text{for} \quad t > 3 \qquad (P_A = 0 \quad \text{when} \quad t < 3),$$

where t is the smaller of τ_{ms}, the main-sequence lifetime of a star, and T, the age of the Galaxy, both measured in units of 10^9 years. The assumption is made here that planetary systems are formed concurrently with their primary stars and, hence, will be of approximately the same age. The maximum value of P_A is reached for stars having τ_{ms} greater than or equal to T.

Based on different pieces of evidence and different interpretations of the observational data, current estimates of T, the age of the Galaxy, range from 10 to 25 billion years. If we take T as 10×10^9 years for our present computations so as not to overestimate the probabilities, then P_A can range between 0 and 0.7 and takes on the values given in Table 16 for stars of each spectral class.

Table 16. Probability That Age Is Greater Than 3 Billion Years

Spectral class	Mean mass	Residence time in main sequence, τ_{ms} (10^9 years)	P_A
F0	1.54	2.12	0
1	1.47	2.66	0
2	1.40	3.47	0.1355
3	1.34	3.98	0.246
4	1.28	4.90	0.388
5	1.24	5.37	0.441
6	1.18	6.46	0.536
7	1.14	7.25	0.586
8	1.10	8.32	0.640
9	1.06	9.34	0.679
G0	1.02	10.2	0.700
1	0.985	11.9	0.700
2	0.955	>12	0.700

PROBABILITY THAT, ALL ASTRONOMICAL CONDITIONS BEING PROPER, LIFE HAS DEVELOPED ON THE PLANET

As we have seen, free oxygen in a planetary atmosphere is essential for a planet to be considered habitable. It is highly probable that virtually all the free oxygen in the Earth's atmosphere has come from the decomposition of water by green plants during photosynthesis. Thus the question becomes: what is the likelihood that this or a similar process would develop on other planets having all of the essential astronomical conditions?

We do not actually know how life started on the Earth, although this topic has been much analyzed in recent years. If we assume that the origin of life has been a natural evolutionary process, however, there is no reason to suppose that life would not always originate whenever the conditions were correct. In support of this view is the evidence that microscopic forms of life appeared on the Earth rather soon after the Earth's formation. The oldest rocks, on the basis of analyses for radioactive decay products, are estimated to be about 4.5 billion years old, while the earliest detectable life forms appear in rocks that are about 2.5 billion years old, or even older in the opinion of some investigators, including T. C. Hoering of the Carnegie Institution's Geophysical Laboratory (Simons, 1962). Since the appearance of detectable life forms must have been preceded by a long period of evolution of chemical precursors and life forms that could leave no permanent evidence of their presence on the Earth, it seems that life originated on the Earth very soon after the environmental conditions became suitable.

Since living matter as we know it is composed of some of the most abundant elements in the universe and in the Earth's crust, there is no reason to suppose that life elsewhere on habitable planets, although undoubtedly differing from Earth life in fine structure, would be greatly different in general chemical composition. Wherever they are found, living organisms must always depend on the same basic chemical processes and physical laws with which we are familiar on the Earth's surface. If water is the only major source of hydrogen available, then living material, wherever it exists, will depend on the development of some process for extracting hydrogen from water and incidentally releasing oxygen to the atmosphere.

Although we are concerned here with the occurrence of life on planets suitable for human beings, many life forms can survive, and probably could originate, under conditions not tolerable to human beings. Thus life-bearing planets are undoubtedly much more numerous in the Galaxy than are planets habitable by man.

It will be assumed here that life will *always* appear on planets having the correct combination of astronomical conditions and that free oxygen in the atmosphere will always accompany the appearance of life.

Whether or not chlorophyll (magnesium-porphyrin protein), for instance, would be found on other planets as the main agent of photosynthesis is an interesting question. Apparently chlorophyll, in various forms designated as chlorophyll-a, -b, -c, -d, and bacteriochlorophyll, has been the undisputed champion among photosynthetic compounds on the Earth for 1½ to 2½ billion years, although other compounds apparently capable of carrying out much the same role, specifically the carotenoids (Fruton and Simmonds, 1959) are employed in conjunction with chlorophyll in various plant species. It is quite possible that life, wherever it appears, may depend on very similar organic compounds for the following reasons. Water (H_2O), ammonia (NH_3), and methane (CH_4) should be among the most abundant compounds in the primitive atmospheres of terrestrial planets in the early stages of their development. A great variety of more complicated organic compounds including the amino acids, other organic acids, pyrroles, purines, and pyrimidines, are formed when mixtures of methane, ammonia, water, and hydrogen are acted upon by electric discharges or by various kinds of radiation (Miller, 1955; Palm and Calvin, 1959; Berger, 1961). These organic compounds would be available as building blocks in living processes (once these had begun); consequently, the most useful products formed in the process would tend to be much the same on every planet. Since chlorophyll and its variants are so very useful and vital on the Earth, perhaps very similar compounds would inevitably evolve elsewhere in the universe through mutation and selection.

For present purposes, then, it will be assumed that P_L is equal to 1.0.

We are here depending on indigenous life to provide an oxygen-rich atmosphere. It does not necessarily follow that any of the indigenous plant life would be edible or palatable from the standpoint of man or animals that have evolved on the Earth. Human settlers on planets beyond the solar system should be prepared to take their own seeds, soil bacteria, *et cetera*, to start crops of the food plants familiar on Earth. Incidentally, it is quite probable that life forms that evolved on the Earth will encounter no natural pathogens or parasites on an alien planet, since parasites and hosts typically must evolve together.

The probability that indigenous *intelligent* life will be present on a given planet seems to be quite low. This is the conclusion of Beadle (1960) based on the enormous number of possible paths that evolution could take. He indicates that the probability of an organism evolving with a nervous system like man's is extremely small. On the Earth, it has been only within the past 100,000 years or so (out of perhaps 2 to 3 billion years during which living things have existed) that the presence of an intelligent species would have been apparent to a visitor from some other planet. It may be that, given enough time, an intelligent species will eventually appear on a given habitable planet. However, its time of appearance is probably not highly predictable. If, through some unlucky accident, the ancestral family group of the human race had become extinct so that there were now no human beings on Earth, how long might it be before some other intelligent species would evolve from the existing animal species of the Earth? The answer is by no means obvious.

On the other hand, once an intelligent species has evolved on a planet, there is a good probability that it will quickly spread to other unoccupied habitable planets in its region of space, as the human race may well do within the span of not too many generations.

The probability of finding a habitable planet already in the possession of an intelligent species is one of the interesting question marks of the future. For present purposes, however, this probability will be regarded as too remote to be considered in making our calculation.

PROBABILITY THAT A GIVEN STAR WILL HAVE AT LEAST ONE HABITABLE PLANET IN ORBIT AROUND IT

Having computed all the essential probabilities for all primaries in the stellar mass range of interest, we have completed all preliminaries to calculating the probabilities that a given star will have habitable planets in orbit about it and the total number of habitable planets in the Galaxy (see page 82 for definitions):

$$P_{HP} = P_p P_i P_D P_M P_e P_B P_R P_A P_L$$

and

$$N_{HP} = N_s P_{HP}.$$

From previous sections we have seen that P_D and P_A depend on the properties of the primary star, and we have adopted the following values: $P_p = 1.0$, $P_i = 0.81$, $P_M = 0.19$, $P_e = 0.94$, $P_R = 0.9$, $P_L = 1.0$, and $P_B = 0.95$ for a star taken at random, but $P_B = 1.0$ if we know that there is no interference from a second star in a binary star system. Thus, for isolated stars, for stars in a widely separated binary system, and for very close (spectroscopic) binaries, $P_{HP} = 0.130 P_D P_A$; and for stars in general, $P_{HP} = 0.124 P_D P_A$. These products are given in Table 17 and P_{HP} for isolated stars are shown in Figure 39. From Table 17 it may be seen that every star in the mass range 0.9 to 1.02 solar masses has about a 5.4 per cent chance of having a habitable planet in orbit around it. In this mass range about one out of eighteen stars should have a habitable planet. More massive stars have a lower probability of having a habitable planet because of a reduced P_A due to their shorter residence times in the main sequence. Less massive stars have a lower probability because of the incompatibility between illumination level and tidal retardation of planetary

Table 17. Probability of a Habitable Planet

Spectral class	Mean mass	P_A	P_D	$P_A P_D$	P_{HP}	
					isolated[a]	general[b]
F0	1.54	0	0.63	0	0	0
1	1.47	0	0.63	0	0	0
2	1.40	0.1355	0.63	0.086	0.0111	0.0106
3	1.34	0.246	0.63	0.1547	0.0202	0.0192
4	1.28	0.388	0.63	0.2442	0.0318	0.0303
5	1.24	0.441	0.63	0.278	0.0362	0.0344
6	1.18	0.536	0.63	0.3374	0.0440	0.0418
7	1.14	0.586	0.63	0.369	0.0481	0.0457
8	1.10	0.640	0.63	0.4025	0.0525	0.0499
9	1.06	0.679	0.63	0.4265	0.0556	0.0529
G0	1.02	0.700	0.63	0.441	0.0574	0.0545
1	0.985	0.700	0.63	0.441	0.0574	0.0545
2	0.955	0.700	0.63	0.441	0.0574	0.0545
3	0.930	0.700	0.63	0.441	0.0574	0.0545
4	0.900	0.700	0.63	0.441	0.0574	0.0545
5	0.870	0.700	0.60	0.420	0.0546	0.0520
6	0.850	0.700	0.59	0.413	0.0538	0.0511
7	0.825	0.700	0.455	0.3185	0.0415	0.0394
8	0.800	0.700	0.36	0.252	0.0328	0.0312
9	0.775	0.700	0.255	0.1784	0.0232	0.0221
K0	0.750	0.700	0.17	0.1190	0.0155	0.0147
1	0.730	0.700	0.07	0.049	0.00638	0.00606
2	0.705	0.700	0	0	0	0

[a] P_{HP} (isolated) $= P_A P_D P_p P_i P_M P_e P_R P_L$.
[b] P_{HP} (general) $= P_A P_D P_p P_i P_M P_e P_R P_L P_B$.

rotation. As was mentioned earlier, stars in the mass range 0.35 to 0.72 solar mass could have habitable planets, but only under the unusual circumstance that they also have a massive close satellite that maintains their rotation rate.

The probability that a star will have two habitable planets in separate orbits around it is quite small for stars with complete ecospheres—about 0.13 per cent. This probability decreases rapidly, however, for stars of less than 0.88 solar mass.

The probabilities of paired habitable planets and of two pairs of paired habitable planets in orbit around a single star are extremely small, but such configurations are not impossible. Four habitable planets (two pairs) around a single star is probably the maximum possible number, but this phenomenon is undoubtedly very rare.

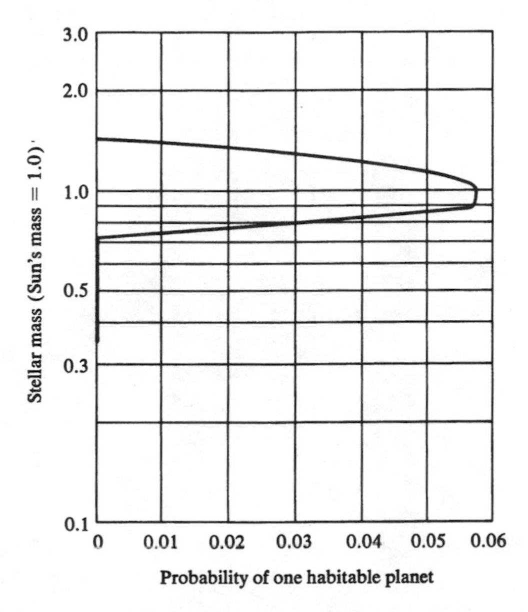

Figure 39. The probability of at least one habitable planet in orbit around an isolated star as a function of stellar mass.

As shown in Table 18, we have estimated that there are about 4.0×10^{-4} habitable planets per cubic parsec of space in our Galaxy, or one habitable planet in every 2480 cubic parsecs, if the sampling of stars in the visible portion of the Galaxy is taken as representative of the whole. Actually we know that the visible part of our Galaxy is much less densely populated with stars than the central part, most of which is obscured by dust. On the other hand, star densities farther toward the rim are somewhat lower than in the solar neighborhood. Since the Galaxy has a volume of about 1.6×10^{12} cubic parsecs, this means that the total number of habitable planets in the Galaxy is about 600 million. This is an estimate of the number of habitable planets in our Galaxy only. There are countless millions of galaxies in the universe, and each is expected to have its own quota of habitable planets.

What does this imply with respect to the number of habitable planets

Table 18. Number of Stars with Habitable Planets per Cubic Parsec of Our Galaxy

Spectral class	Number of stars per cubic parsec	P_{HP}, general	Number of stars with habitable planets per cubic parsec
F2	2.76×10^{-4}	0.0106	2.93×10^{-6}
3	3.20×10^{-4}	0.0192	6.15×10^{-6}
4	3.66×10^{-4}	0.0303	1.11×10^{-5}
5	3.92×10^{-4}	0.0344	1.35×10^{-5}
6	4.04×10^{-4}	0.0418	1.69×10^{-5}
7	4.12×10^{-4}	0.0457	1.88×10^{-5}
8	4.25×10^{-4}	0.0499	2.12×10^{-5}
9	4.36×10^{-4}	0.0529	2.31×10^{-5}
G0	4.43×10^{-4}	0.0545	2.41×10^{-5}
1	4.52×10^{-4}	0.0545	2.46×10^{-5}
2	4.55×10^{-4}	0.0545	2.48×10^{-5}
3	4.99×10^{-4}	0.0545	2.72×10^{-5}
4	5.35×10^{-4}	0.0545	2.92×10^{-5}
5	6.40×10^{-4}	0.0520	3.33×10^{-5}
6	6.90×10^{-4}	0.0511	3.53×10^{-5}
7	7.60×10^{-4}	0.0394	2.99×10^{-5}
8	7.90×10^{-4}	0.0312	2.46×10^{-5}
9	8.35×10^{-4}	0.0221	1.84×10^{-5}
K0	8.74×10^{-4}	0.0147	1.28×10^{-5}
1	9.14×10^{-4}	0.00606	5.54×10^{-6}
	Total 1.09×10^{-2}		Total 4.03×10^{-4}

NOTES: Total of all stars = 8.52×10^{-2} per cubic parsec.
About 0.47 per cent of all stars have habitable planets.
About 3.7 per cent of stars of classes F2 to K1 have habitable planets.
Volume of Galaxy is 1.6×10^{12} cubic parsecs.
Number of habitable planets in Galaxy is 645 million.

in the local region of our Galaxy within the galactic spiral arm in the vicinity of the Sun? It is a bit difficult to visualize a cubic parsec; therefore Table 19 is included to give the expected number of habitable planets that should be found within a spherical volume of radius R centered at the Sun.

Within 100 light-years of the Earth (a relatively small distance when it is considered that the thickness of our Galaxy is over 10,000 light-years at the center and its diameter is about 80,000 light-years), there should be about 50 habitable planets.

Table 19. Expected Number of Habitable Planets N within a Sphere of Radius R from the Sun

R (light-years)	N
27.2	1
34.3	2
46.5	5
58.5	10
100.0	50

Throughout the Galaxy, the mean distance between a given star chosen at random and its closest stellar neighbor is about 4 light-years; the mean distance between a star with a habitable planet and its closest neighbor with a habitable planet is about 24 light-years.

In general, if $\bar{\rho}$ is the mean density of a given class of objects, the mean distance $\bar{r}$ between a given object and its closest neighbor may be expressed as

$$\bar{r} = \frac{\Gamma(\frac{4}{3})}{(\frac{4}{3}\pi\bar{\rho})^{\frac{1}{3}}} = 0.554(\bar{\rho})^{-\frac{1}{3}}.$$

In the central regions of globular clusters, mean separations between neighboring stars may be measured in light-days or light-weeks rather than in light-years (1 light-week is equal to approximately 1000 astronomical units). Therefore, mean separations between habitable planets (if planets can exist on stable orbits in these densely populated regions) would possibly be measured in thousands of astronomical units.

CHAPTER 6

The Nearest Candidates

Now that the general characteristics of stars that could have habitable planets have been determined, it is possible to consider the individual stars in the immediate vicinity of the Sun and to make an estimate of the probabilities that these stars have habitable planets in orbit about them.

Of the one hundred nearest stars (plus some eleven unseen companions) listed in Allen's table (Allen, 1955) of the stars within 22 light-years from the Sun, approximately forty-three could have habitable planets (see Table 21). All but fifteen of these are so small, however, that they could have a habitable planet only if the planet also possessed a satellite large enough and close enough to maintain its rotation rate. Hence, the proportion of stars that have a reasonable likelihood of having habitable planets in the solar neighborhood is of the order of 13 per cent. The remaining sixty-eight stars were omitted for the following reasons: Three (Sirius, Procyon, and Altair) are excessively massive and therefore too short-lived; seven are white dwarfs; fifty-seven are too small, and they would either retard planetary rotation at ecosphere distances or produce destructive tides on those planets that had their rotation maintained by a large, close satellite; one (40 Eridani A), otherwise acceptable, is in a system with a nearby white dwarf.

One of the stars listed in Tables 20 and 21 (Lal 21185 A) should, perhaps, also be omitted as a candidate. It is in a multiple star system of which the orbital characteristics (although not well established) would, if present calculations are correct, be incompatible with the existence of stable planetary orbits within an ecosphere. The characteristics of the multiple star systems of Table 21 are given in Table 20.

Table 20. Characteristics of Multiple Star Systems in Table 21

System	Period (years)	Orbital eccentricity	Semi-major axis (A.U.)	Total mass	Mass of component A	Mass of component B	Present separation (A.U.)	Absolute visual magnitude, M_v, component B	Remarks
α Centauri	80.09	0.52	23.2	1.96	1.08	0.88	...	...	Component C, 2° removed, has an absolute visual magnitude M_v of 15.4.
61 Cygni	697 (?) to 783 (?)	0.40	83 to 112	1.14	0.63	0.51	...	...	Component C may have mass as low as 0.008 solar mass.
70 Ophiuchi	87.8	0.50	22.8 to 24.3	1.55	0.90	0.65	...	...	Third component has not been verified.
η Cassiopeiae	480 (?) to 526 (?)	0.53	69 to 71	1.52	0.94	0.58	...	...	Third component is not well established.
p Eridani	251	...	52 to 56	2.18	1.09	1.09	...	...	Masses given here appear to be too high to be compatible with absolute magnitudes.
Lal 21185	1.28 (?) to 8 (?)	...	...	...	...	$\sim$0.02	...	...	Orbital characteristics may prevent the occurrence of a habitable planet.
Grm 34	...	...	...	...	...	...	$>$136	13.45	Component A is a spectroscopic binary; component B is of spectral class M5.
36 Ophiuchi	...	...	...	...	...	...	$\sim$22	...	A three-component system; component C is 12 minutes of arc away from A–B.
HR 7703	...	...	...	...	...	...	$\sim$46	12.7	Component B is of spectral class M5.
HR 5568	...	...	...	...	...	...	$>$57	...	...
+44° 2051	...	...	...	...	...	...	$\sim$165	14.2 to 16.0	Component B is of spectral class M7e.
HR 753	...	...	...	...	...	...	$>$1100	12.58	Component B is of spectral class M6.

Table 21. The Stars within 22 Light-years

Name of star	Other designations	Right ascension, RA, 1900		Declination, 1900		Apparent visual magnitude	
		(hours)	(minutes)	(degrees)	(minutes)	Allen	Boss
α Centauri A	Rigel Kentaurus	14	32.8	−60	25	0.02[a]	0.33
α Centauri B		14	32.8	−60	25	1.39[a]	1.70
Lal 21185 (A)	BD + 36° 2147	10	57.9	+36	38	7.54	7.60
ε Eridani		3	28.2	−9	48	4.2	3.81
61 Cygni A		21	2.4	+38	15	5.29	5.57
61 Cygni B		21	2.4	+38	15	6.06	6.28
ε Indi		21	55.8	−57	12	4.7	4.74
Grm 34 A	BD + 43° 44	0	12.7	+43	27	8.18	8.1
Lac 9352	CD − 36° 15693	22	59.4	−36	26	7.2	7.44
τ Ceti		1	39.4	−16	28	3.65	3.65
Lac 8760	CD − 39° 14192	21	11.4	−39	15	6.65	6.65
Cin 3161	CD − 37° 15492	23	59.5	−37	51	8.6	8.57
Grm 1618	BD + 50° 1725	10	5.3	+49	57	6.75	6.82
CC 1290	− 49° 13515	21	26.9	−49	26	8.6	...
Cin 18,2354	+ 68° 946	17	37.0	+68	26	9.15	9.5
+15° 2620	HD 119850	13	40.7	+15	26	8.58	8.5
70 Ophiuchi A		18	0.4	+2	31	4.19	4.23
70 Ophiuchi B		18	0.4	+2	31	5.87	6.23
η Cassiopeiae A		0	43.0	+57	17	3.54	3.64
η Cassiopeiae B		0	43.0	+57	17	7.4	...
σ Draconis		19	32.6	+69	29	4.72	4.78
36 Ophiuchi A		17	9.2	−26	27	5.17	...
36 Ophiuchi B		17	9.2	−26	27	5.20	...
36 Ophiuchi C		17	9.2	−26	27	6.53	...
HR 7703 A	HD 191408	20	4.6	−36	21	5.24	5.34
HR 5568 A	HD 131977	14	51.6	−20	58	5.90	5.76
HR 5568 B	HD 131976	14	51.6	−20	58	8.08	8.87
δ Pavonis		19	58.9	−66	26	3.67	3.64
−21° 1377		6	6.4	−21	49	8.3	...
+44° 2051 A	Lal 21258	11	0.5	+44	2	8.7	8.8
+4° 4048 (A)		19	12.1	+5	2	9.18	...
HD 36395	−3° 1123	5	26.4	−3	42	7.96	8.4
+1° 4774	Lal 46650	23	44.0	+1	52	9.05	8.8
+53° 1320		9	7.6	+53	7	7.90	...
+53° 1321		9	7.6	+53	7	8.01	...
−45° 13677		20	6.7	−45	28	8.4	...
82 Eridani		3	15.9	−43	27	4.3	4.30
β Hydri		0	20.5	−77	49	2.9	2.90
HR 8832		23	8.5	+56	37	5.67	5.65
+15° 4733		22	51.8	+16	2	8.69	...
p Eridani A		1	36.0	−56	42	6.1	6.00
p Eridani B		1	36.0	−56	42	6.1	6.03
HR 753 A		2	30.6	+6	25	5.94	5.92

Note: Lal—*Lalande's Star Catalogue* (1837); BD—*Bonner Durchmusterung*; Grm—*Cordoba Durchmusterung* (1886); HD—*Henry Draper Catalogue* (1918–1924); HR—

[a] van de Kamp (1958) gives $m_{v_A} = 0.09$, $m_{v_B} = 1.38$.

[b] Very small; less than 0.001.

That Could Have Habitable Planets

Parallax, π (sec)		Spectral class, Allen	Distance (light-years), Allen	Absolute visual magnitude, M_v, Allen	Adopted mass	Probability of habitable planet, P_{HP}	
Allen	Boss						
0.754	0.756	G4	4.3	4.5	1.08	0.054	0.107
0.754	0.760	K1	4.3	5.9	0.88	0.057	
0.398	0.388	M2	8.2	10.51	0.37	(*b*)	
0.303	0.305	K2	10.8	6.2	0.80	0.033	
0.293	0.299	K5	11.1	7.65	0.63	(*b*)	
0.293	0.299	K8	11.1	8.42	0.51	(*b*)	
0.288	0.288	K5	11.3	7.0	0.71	(*b*)	
0.278	0.284	M2	11.7	10.44	0.38	(*b*)	
0.273	0.278	M1	12.0	9.4	0.47	(*b*)	
0.268	0.301	G8	12.2	6.02	0.82	0.036	
0.258	0.257	M0	12.6	8.7	0.54	(*b*)	
0.219	0.222	M3	14.9	10.3	0.39	(*b*)	
0.219	0.218	K8	14.9	8.45	0.56	(*b*)	
0.212	...	M3	15.4	10.5	0.37	(*b*)	
0.203	0.212	M3	16.1	10.75	0.35	(*b*)	
0.193	0.191	M1	16.9	10.0	0.42	(*b*)	
0.188	0.196	K1	17.3	5.7	0.90	0.057	
0.188	0.196	K5	17.3	7.3	0.65	(*b*)	
0.181	0.182	F9	18.0	4.87	0.94	0.057	
0.181	...	K6	18.0	8.7	0.58	(*b*)	
0.179	0.181	G9	18.2	6.01	0.82	0.036	
0.179	0.178	K2	18.2	6.4	0.77	0.023	0.042
0.179	0.178	K1	18.2	6.5	0.76	0.0195	
0.179	0.178	K6	18.2	7.8	0.63	(*b*)	
0.175	0.178	K2	18.6	6.5	0.76	0.0195	
0.174	0.172	K4	18.8	7.1	0.70	(*b*)	
0.174	0.172	M0	18.8	9.2	0.50	(*b*)	
0.170	0.155	G7	19.2	4.9	0.98	0.057	
0.170	...	M0	19.2	9.6	0.455	(*b*)	
0.170	0.175	M0	19.2	9.9	0.43	(*b*)	
0.168	...	M3	19.4	10.33	0.39	(*b*)	
0.163	0.168	M1	20.0	9.06	0.51	(*b*)	
0.161	0.164	M2	20.2	10.18	0.40	(*b*)	
0.161	...	K7	20.2	8.9	0.52	(*b*)	
0.161	...	K9	20.2	9.0	0.51	(*b*)	
0.158	...	M0	20.6	9.4	0.48	(*b*)	
0.156	0.159	G5	20.9	5.3	0.91	0.057	
0.153	0.144	G1	21.3	3.8	1.23	0.037	
0.152	0.146	K3	21.4	6.69	0.74	0.011	
0.150	...	M2	21.8	9.72	0.445	(*b*)	
0.148	0.163	K2	22.0	7.0	0.71	(*b*)	
0.148	0.163	K2	22.0	7.1	0.70	(*b*)	
0.148	0.144	K3	22.0	6.79	0.725	0.004	
						Net 0.434	

Groombridge's Catalogue of Circumpolar Stars; Lac—*Lacaille's Catalogue* (1847); CD—*Revised Harvard Photometry* (1908).

The fourteen most promising candidates, with probabilities of possessing a habitable planet exceeding 1 per cent, are listed in Table 22 in the order of their distances from the Earth. The probability that there is at least one habitable planet among these fourteen stars is 0.43. A sky map showing their locations is given as Figure 40.

Table 22. Fourteen Stars Most Likely To Have Habitable Planets in Order of Distance from Earth

Star	Distance from Earth (light-years)	Probability, P_{HP}	
Alpha Centauri A	4.3	0.054	0.107
Alpha Centauri B	4.3	0.057	
Epsilon Eridani	10.8	0.033	
Tau Ceti	12.2	0.036	
70 Ophiuchi A	17.3	0.057	
Eta Cassiopeiae A	18.0	0.057	
Sigma Draconis	18.2	0.036	
36 Ophiuchi A	18.2	0.023	0.042
36 Ophiuchi B	18.2	0.020	
HR 7703 A	18.6	0.020	
Delta Pavonis	19.2	0.057	
82 Eridani	20.9	0.057	
Beta Hydri	21.3	0.037	
HR 8832	21.4	0.011	

No list of the nature of Table 22 can be absolutely complete in all respects, since the pertinent parameters are actually not known with the precision that is implied in the published data. Taking the data at face value, but recognizing that they are subject to future revisions, the probabilities for the existence of a habitable planet in orbit about each star are given. Where no determinations of the masses of the stars were available, masses were estimated from the absolute magnitudes. In those cases where binary stars had characteristics such that either star would be a suitable primary, probabilities were calculated also for the existence of at least one habitable planet in the binary system. For example, in the case of the nearest stellar system, Alpha Centauri, the individual probabilities are 0.054 and 0.057 for components A and B, respectively; for the system, the probability is 0.107; that is, there is about a 10 per cent

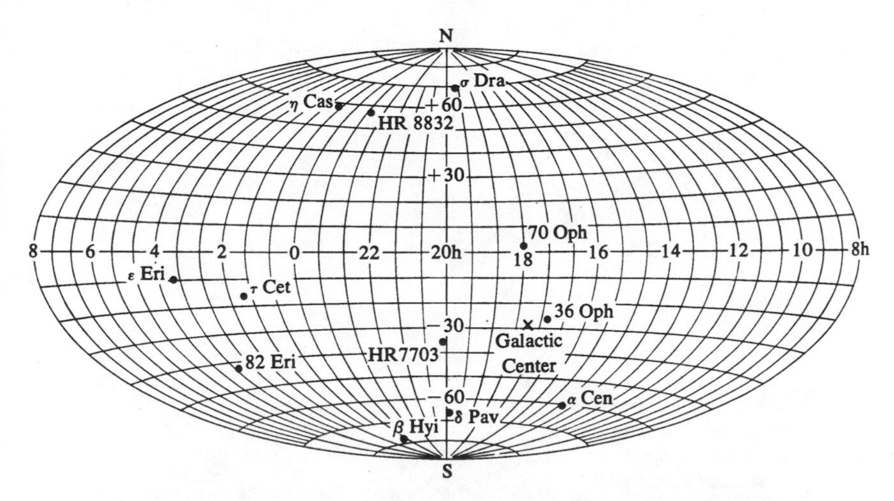

Figure 40. Positions of the nearest stars that have significant probabilities of possessing at least one habitable planet; coordinates of right ascension and declination.

chance of the existence of a habitable planet in the Alpha Centauri system.

Because of its location close to the south celestial pole, Alpha Centauri can not be seen from positions on the Earth's surface north of latitude 30°N (roughly the latitude of New Orleans). The apparent orbit of component B around component A, as obtained from numerous telescopic observations over the past 100 years, is shown in Figure 41. It is an extremely elongated orbit, since it is seen almost edge-on from the Earth. The actual orbit has an eccentricity of 0.52 and a semimajor axis of 23.2 astronomical units; hence, at their closest approach (periastron), A and B are separated by 11.2 astronomical units while their separation at apastron (greatest separation) is 35.3 astronomical units. The approximate distances at which ecospheres exist around A and B are shown to scale in the figure. Although there is no theoretical method of determining the stability of orbits of planetary bodies in multibody systems, application of the stability limits obtained from the restricted three-body problem suggests that habitable planets (if any) orbiting within the ecospheres of A and B should have highly stable orbits.

For example, if A and B were on circular orbits around their common center of gravity, taking their mass ratio as 0.45 and placing them at such a distance from each other (9.73 astronomical units) that their new angular velocity was equal to their actual angular velocity at periastron, then planetary orbits of A should be stable within a radius of 2.68 astronomical units, and planetary orbits of B should be stable within a radius of 2.34 astronomical units. From Figure 26 (see page 72), it may be seen that complete ecospheres of both A and B fall well inside these stability boundaries.

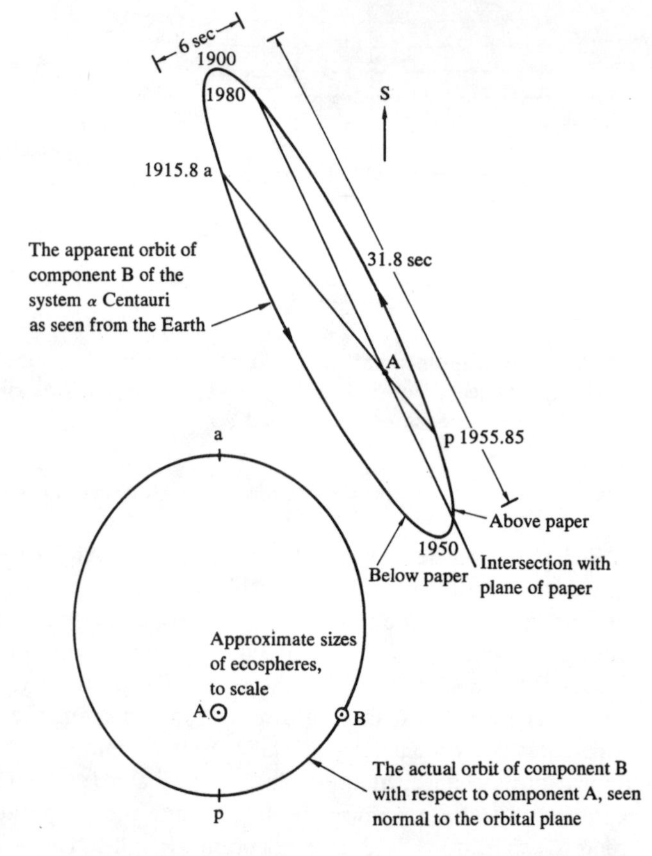

Figure 41. The orbit of component B of the system Alpha Centauri.

The larger component of the system, α Centauri A, is a star very similar to the Sun. Its spectral class is given as G4 (or sometimes as G0); its apparent visual magnitude, as 0.09; its absolute visual magnitude, as 4.5; and its mass, as about 1.08 solar masses. It has a probability of possessing a habitable planet of approximately 0.054.

Component B is somewhat smaller, of spectral class K1 (or K5); it has an apparent visual magnitude of 1.38, an absolute visual magnitude of 5.9 (or 6.1), and a mass of 0.88 solar mass. Thus its probability of possessing one habitable planet is 0.057. Component C, also called Proxima

Centauri, is 2.2 degrees away from the other two, as seen from the Earth. It is a small flare star of visual magnitude 10.68, too small to be considered the center of a system in which a habitable planet could be found.

The probabilities are extremely sensitive to the values of stellar mass derived from the astronomical data. It is clear that none of the values of stellar mass that have been determined so far is very precise. As better data are obtained, future revisions to the mass determinations will necessitate some changes in the probability figures given.

Of the remaining stars in the list of the most promising candidates, relevant information is given below.

Epsilon Eridani, located in the sky at almost 10 degrees south of the projected plane of the Earth's equator (declination, 9° 48′), can be seen from any part of the Earth's surface, except from a small region around the North Pole. It is an isolated star (no companion has ever been detected) of spectral class K2, although it is sometimes classed as K0. Its apparent visual magnitude is 4.2, its parallax is 0.303 second, and its distance from the Earth is 10.8 light-years. Therefore, its absolute visual magnitude is 6.2. From this it may be deduced that its mass, which can not be measured directly, is about 0.80 the mass of the Sun. The probability that there is a habitable planet orbiting within its ecosphere is here calculated to be 3.3 per cent.

Tau Ceti, fairly close to ε Eridani in the night sky (in a neighboring constellation), can be seen from any point on the Earth's surface, except from the arctic regions. Both ε Eridani and τ Ceti were recently "listened to" by means of a radio telescope during the course of Project Ozma, an attempt to detect intelligence-bearing radio signals directed toward our Sun by possible intelligent inhabitants of planets of these stars. This was based on a suggestion of Cocconi and Morrison (1959) of Cornell University, who selected a frequency of 1420 megacycles (wave length, 21 centimeters) as the optimum for a high signal-to-noise ratio. An attempt to detect radio signals from ε Eridani and τ Ceti was made in the spring of 1960 by using the 85-foot diameter radio telescope of the National Radio Astronomy Observatory at Green Bank, West Virginia. Results were negative. That no signals were detected is not surprising, since the reverse experiment conducted on planets orbiting around ε Eridani or τ Ceti but directed toward our Sun would also have given negative results. We are not sending out signals on a wave length of 21 centimeters that "they" could detect. The joint probability that ε Eridani and τ Ceti have one habitable planet between them is only about 7 per cent, while the probability that a given habitable planet will be inhabited by intelligent beings is difficult to estimate.

Like ε Eridani, τ Ceti is apparently an isolated star. Its spectral class is variously given as G8, G4, and K0; its absolute visual magnitude, as 6.02, 5.8, and 5.9; and its distance, as 12.2, 11.8, and 11.2 light-years. Its apparent visual magnitude is 3.65. Based on an absolute visual magnitude of 6.02, its mass is estimated to be 0.82 solar mass. The probability that it has one habitable planet in orbit about it is here calculated to be 3.6 per cent.

70 *Ophiuchi A* is the more massive component of the system 70 Ophiuchi. This system consists of two stars revolving about each other with a period of 87.85 years in an orbit with an eccentricity of 0.50. Its distance from the Earth is given variously as 17.3, 16.4, and 16.5 light-years. The parallax given by van de Kamp (1958), $\pi = 0.199$, corresponds to 16.4 light-years. The semimajor axis of the binary orbit is 22.8 astronomical units; thus at periastron the two components are separated by 11.4 astronomical units; at apastron they are 34.2 astronomical units apart. No third companion has been established for the 70 Ophiuchi system, although dark companions are suspected to exist. The apparent visual magnitude is 4.19. Component A of spectral class K1 has an absolute visual magnitude of 5.7 (or 5.8), a mass of about 0.90 solar mass, and hence a 5.7 per cent probability of possessing one habitable planet. According to van de Kamp (1958), the spectral class of A is K0, and its apparent visual magnitude is 5.09.

The less massive component B is of spectral class K5 with an absolute visual magnitude of 7.3 (or 7.4 or 7.5). Its mass is about 0.65 solar mass; thus it could possess a habitable planet only if it had a large, close satellite to preserve its rotation rate. van de Kamp gives the spectral class of B as K4. Planets revolving at ecosphere distance should have stable orbits in this binary system.

As seen from the system of 70 Ophiuchi, the Sun would appear as a third-magnitude star in the constellation Orion, not far from the belt.

Eta Cassiopeiae A is the larger component of the binary system η Cassiopeiae. This system, at a distance of 18.0 light-years from the Earth, has a period of the order of 500 years and an orbital eccentricity of 0.53. The semimajor axis is about 70 astronomical units. The existence of a third component is not well established. Apparent visual magnitude is 3.54. Component A is of spectral class F9, has an absolute visual magnitude of 4.87 (very close to that of the Sun), and a mass of about 0.94 solar mass. Its probability of having one habitable planet is 5.7 per cent.

The smaller component B is of class K6, absolute visual magnitude 8.7, and mass 0.58 solar mass; thus its probability of having a habitable planet is very small.

Our Sun, as seen from this system, would appear to be imbedded in the Southern Cross.

Sigma Draconis, the most northerly star in this list (declination, 69° 29′), appears to be an isolated star. At a distance of 18.2 light-years, with an apparent visual magnitude of 4.72, it is in spectral class G9. Its absolute visual magnitude is 6.01, and its mass is about 0.82 solar mass. There is a probability of approximately 3.6 per cent that it has a habitable planet.

36 *Ophiuchi A* is the most massive member of a system that lies almost directly between us and the center of our Galaxy. Apparently its orbital elements have not yet been established. According to Allen (1955), A and B are separated by 4 seconds of arc, while component C is over 12 minutes of arc away from A–B. The system is about 18.2 light-years away and has an apparent visual magnitude of 5.17. Component A, in spectral class K2, has an absolute visual magnitude of 6.4 and a mass of about 0.77 solar mass. Thus it has a probability of about 2.3 per cent of having a habitable planet. Component B, in spectral class K1, has an absolute visual magnitude of 6.5 and a mass of about 0.76 solar mass. Its probability of having a habitable planet is 0.020. Component C, in spectral class K6, has an absolute visual magnitude of 7.8 and a mass of about 0.63. The probability of its having a habitable planet is very small.

HR 7703 A is the larger member of its system, which is 18.6 light-years away in the southern constellation Sagittarius and consists of two stars separated by 8 seconds of arc. The orbital elements have not yet been determined. Apparent visual magnitude is 5.24. Component A is of spectral class K2, with absolute visual magnitude of 6.5, estimated mass of 0.76 solar mass, and a 2-per-cent probability of having one habitable planet. The smaller component B is of spectral class M5, has an absolute visual magnitude of 12.7, and is too small to possess a habitable planet.

Delta Pavonis, even more southerly than α Centauri, can not be seen by observers on the Earth's surface north of latitude 23°N. It is apparently an isolated star of spectral class G7. Its apparent visual magnitude is 3.67; its distance from our Sun, 19.2 light-years; its absolute visual magnitude, 4.9; and its mass, 0.98 solar mass. On the basis of this mass value, it has a 5.7-per-cent probability of possessing one habitable planet.

82 Eridani, another apparently isolated star, can be seen from latitudes south of 46°N. Its spectral class is G5; its apparent visual magnitude is 4.3; its distance from our Sun is 20.9 light-years; its absolute visual magnitude is 5.3; and its deduced mass is 0.91 solar mass. Its probability of possessing a habitable planet is 5.7 per cent.

Beta Hydri is the most southerly star on the list of best candidates within 22 light-years from the Sun. It is 21.3 light-years away and is an isolated G1 star with an apparent visual magnitude of 2.90, an absolute visual magnitude of 3.8, and a deduced mass of 1.23 solar masses. It has a 3.7-per-cent probability of possessing a habitable planet.

HR 8832, in the constellation Cassiopeia, is too faint to be seen with the naked eye except under unusually good viewing conditions. It is an isolated star in spectral class K3 with an apparent visual magnitude of 5.67. Located at a distance of 21.4 light-years from the Earth, its absolute visual magnitude is 6.69; its mass is 0.74 solar mass; and its probability of possessing a habitable planet is 1.1 per cent.

The combined probability of the existence of at least one habitable planet in the whole volume of space out to a distance of 22 light-years from the Sun is about 43 per cent.

CHAPTER 7

Star Hopping

CONNING A PLANET

From what distance can it be ascertained that a specific star actually possesses a habitable planet? How closely must we approach a planet in order to be sure that it is habitable?

As we know from experience, there are still many unanswered questions about Venus and Mars, although the Earth periodically comes closer to them than 30 and 40 million miles, respectively. Much of the difficulty lies in the fact that we are looking through our own radiation-absorbing and image-distorting atmosphere. Undoubtedly we will learn a great deal more about the planets of our own system once telescopes of even moderate size are put into operation above the atmosphere.

Imagine, however, that we are on an exploratory mission on a spaceship in space, approaching a star. We have telescopes up to 60 inches in diameter, sensitive radio receivers, and any other necessary sensory equipment. We would like to make a decision at the earliest possible moment whether to continue going toward the star or break off and return to base or head for a second objective. Let us say that our first decision will be based on whether the star has a planet orbiting within an ecosphere. If it has such a planet, we will proceed; if we can say that it definitely does not, we will break off. Subsequent decisions will have to be made if the star does have an orbiting planet; we would like to know the answers to a number of questions relating to the distances at which such decisions might be made. As examples, we might consider the following questions:

1. At what distance could we detect a planet like Jupiter?
2. At what distance could we detect a planet like the Earth?

3. At what distance could the most powerful Earth radio station be detected in space?
4. At what distance could the oceans of Earth be identified?
5. At what distance could forests be identified?
6. At what distance could cities be seen?
7. At what distance could any other works of man be recognized as artifacts?

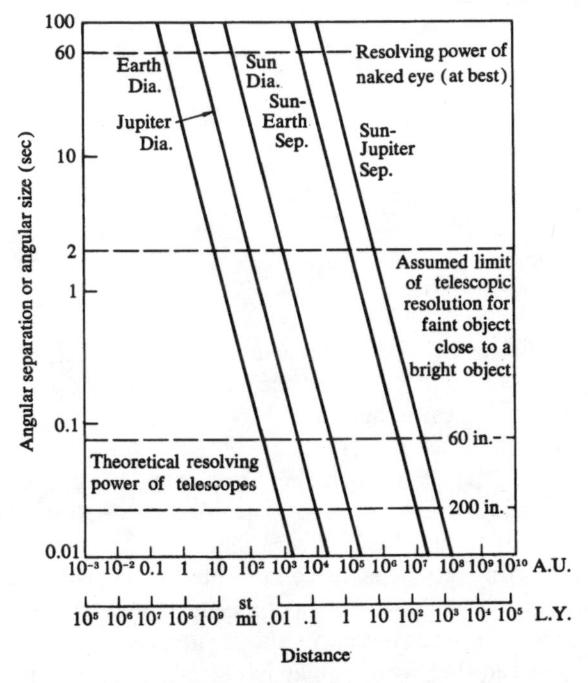

Figure 42. Angular separation or size of objects as a function of distance from observer.

Some of these questions can not be answered definitively at present, but some approximate answers can be suggested. When viewing the universe through our atmosphere, it is difficult for us to observe a faint object when it is close to a much brighter object because the light from the brighter object interferes and because the degraded "seeing" through our atmosphere prevents the realization of the theoretical resolving power of large telescopes. A telescope employed in space, however, free from atmospheric effects, would be able to attain its theoretical resolving power for objects of equal brightness. Also, faint objects close to brighter ones might be observed through use of special techniques such as occulting (hiding) the image of the brighter object with a knife edge. It has been indicated that this technique could be used if the two objects are separated by 2 seconds of arc or more (Roman, 1959). It will be assumed here, then, that a planetary object can be detected if it is separated by at least 2

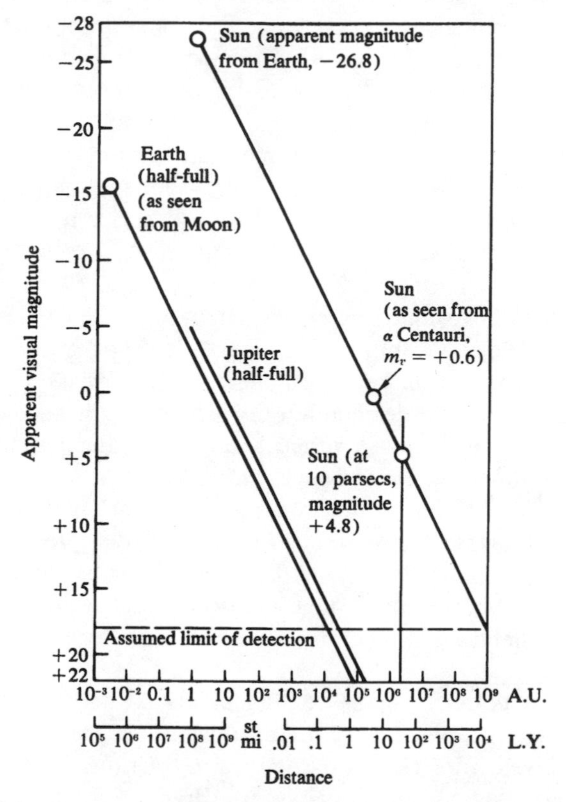

Figure 43. Apparent magnitude of an object as a function of distance from observer.

seconds of arc from its primary and if its apparent visual magnitude is +18 or brighter (roughly the limiting magnitude of the 60-inch telescope).

Under these assumptions, Questions 1 and 2 can be answered by referring to Figures 42 and 43. If we assume that Jupiter is at its greatest elongation and is half-illuminated by the Sun (quarter phase), that it has an albedo of 0.4 and reflects light in a manner similar to a Lambert body, then its absolute visual magnitude would be +27.4. Its maximum elongation from the Sun would be 2 seconds of arc at a distance of 8.55 light-years and its magnitude would be +18 at a distance of 0.43 light-year. Thus the apparent visual magnitude sets the maximum distance for detection. At 0.43 light-year away, the separation at greatest elongation would be about 37 seconds of arc. Its image, however, would be imbedded in a field or background containing so many distant stars of like magnitude that each such object would have to be analyzed spectroscopically to discriminate between stars and planets, or else repeated photographs would have to be taken to detect relative motion. Much would depend on the tilt of the system with respect to the line of approach—the most favorable approach line being at a right angle to the plane of the planetary orbit, such that the planet is always seen at greatest elongation from its primary. Under less favorable conditions of approach, with the planet between us and the star, detections could be made only at much shorter distances. Jupiter's disk, for example, would be discernible as an object 2 seconds in diameter at about 90 astronomical units.

Similar difficulties would prevail in detecting the Earth. Assuming an albedo of 0.36, the applicability of Lambert's law, and greatest elongation, Earth–Sun distance would be resolvable at 1.7 light-years, but the Earth's reflected light would be detectable only when the distance was reduced to 0.17 light-year (10,700 astronomical units); the Earth's disk would be perceptible at about 7.6 astronomical units.

We would not be able to detect the presence of an Earth-like planet, then (under the present assumptions), until we had approached the target system to within $\frac{1}{6}$ of a light-year. Unfavorable indications, such as the presence of a giant planet in such a position that it would prevent the orbiting of a habitable planet within an ecosphere, might be detected earlier.

As has been pointed out by E. Sänger (1962) and others, there is a complication in the optical detection of objects in space when the observer is moving at velocities that are significant fractions of the velocity of light. At nearly relativistic velocities, the light from the star being approached seems to shorten in wave length and becomes more blue; or, in other words, the star appears to be hotter than it actually is. On the other hand, when one is moving away from a star at such velocities, its light seems to shift toward the red end of the spectrum, and the star appears to be cooler than it is. In general, the light coming from the target star will shift in accordance with the relationship

$$\frac{\lambda}{\lambda_0} = \left[\frac{1 - v/c}{1 + v/c}\right]^{\frac{1}{2}},$$

where λ_0 is the original wave length, λ is the perceived wave length, v is the vehicle's velocity, and c is the speed of light.

Similarly, the light from the takeoff star will shift according to the relationship

$$\frac{\lambda}{\lambda_0} = \left[\frac{1 + v/c}{1 - v/c}\right]^{\frac{1}{2}}.$$

At fairly high relativistic velocities, a black spot will appear in the directly forward direction, since all the visually detectable radiation from the stars in that part of the field of view is shifted into the ultraviolet. Another black spot will appear in the directly backward direction, since all the visually detectable radiation from the stars in that sector is shifted into the infrared. Between the two black spots, the stars will range in color from blue to red. In the visual black spots, however, the stars will still be detectable with instruments.

These effects would start to become apparent to the casual observer at velocities of about one-twentieth the speed of light, while the black spots would appear at vehicle velocities of about one-half the speed of light.

In turning next to Question 3 on page 118, radio waves in the "broadcast band" (535 to 1605 kilocycles) originating on Earth do not penetrate the Earth's ionosphere very effectively, and most of the energy is refracted and returned to Earth. It is only the very-high-frequency radio waves (radio waves of frequencies higher than about 30 megacycles) that would be detectable in space. Attenuation and refraction in the ionosphere are most pronounced below approximately 100 megacycles. In the very-high-frequency bands, some of the most powerful Earthly sources are the BMEWS (Ballistic Missile Early Warning System) radar stations broadcasting signals on a frequency of about 400 megacycles. Taking the peak power level as about 1 megawatt and assuming that the natural galactic noise at the receiver is that of a thermal source at 300°K, the signals from BMEWS transmitters might be detected at a distance of 10 billion miles, or 100 astronomical units. (Pluto's mean distance is about 4 billion miles from the Earth.) The BMEWS example is included only for illustration, since, in the search for habitable planets, a strong intelligence-bearing radio signal would be expected to originate only on a planet possessing an intelligent life form. Certain kinds of natural radio noise, however, may be characteristic of planets that, like the Earth, have numerous thunderstorms and radio-static-generating lightning flashes occurring in their atmospheres at all times. It has been estimated that some two to six thousand thunderstorms are in progress on the Earth at any given instant. A recognizable signature of such electrical discharges would at least serve as an indicator of a planet with an atmosphere, but not necessarily a habitable one. What fraction of this noise would leak out and be detectable in space at a distance from the source is not known with any degree of certainty.

Characteristic spectral absorption lines due to oxygen or water vapor might be detected as soon as a planetary image is obtained. The detection of these lines can be accomplished only with great difficulty by men looking through the Earth's atmosphere because the prevalence of oxygen, water vapor, and other absorbing gases in our atmosphere render it opaque in many interesting portions of the electromagnetic spectrum; but such detection would be quite feasible from space.

Estimates of the approximate distances at which important surface features of the Earth could be identified from space (Questions 4 through 7, above) are as follows: oceans at about 7 astronomical units, forests at about 3 astronomical units, and large cities at about 5 million miles with good viewing conditions. Closer approaches and very clear atmospheres would be needed in order to identify as artifacts other conspicuous works of man such as canals, dams, bridges, airfields, and railroads.

Generally speaking, a system must be approached quite closely to make sure that it contains a habitable planet, although negative indications might be obtained while the exploring party is still as far away as $\frac{1}{3}$ to $\frac{1}{2}$ light-year. To be fairly certain that a planet is habitable requires an approach within 1 million miles or so, while to be absolutely certain necessitates a landing on its surface and direct investigations of the planetary atmosphere, surface conditions, and other pertinent parameters.

Any indication that a planet is already inhabited by intelligent creatures would signal the need for proceeding with the utmost caution. In fact, before a manned landing is made on any likely looking planet, it would be desirable to study the planet thoroughly from a distant orbit about it for a protracted period of time, to send sampling probes into its atmosphere, and to send surveillance instruments down to the surface. Contacts with alien intelligence should be made most circumspectly, not only for the protection of mankind or as insurance against unknown factors, but also to avoid any disruptive effects on the local population produced by encountering a vastly different cultural system. After prolonged study of the situation, it would have to be decided whether to make overt contact or to depart without giving the inhabitants any evidence of the visitation

As man's ability to accelerate payloads to higher and higher velocities increases, a point will eventually be reached when interstellar flights can begin to be considered feasible. At the present time, we have barely reached the point in technology where we could design a rocket vehicle capable of sending a small package completely out of the solar system. Its velocity at a distance equivalent to the orbit of Pluto would be very small relative to the speed of light, however, perhaps one ten-thousandth the speed of light. Consequently, trips to even the closest stars would take many thousands of years; and it would not be worth while even to consider such a slow trip. But if one has confidence in man's ability to learn—a a confidence justified by looking back at the accelerating pace of technical progress over the past 400 years—then one may be optimistic about the future feasibility of flights over interstellar distances. Flights at near-relativistic velocities do not violate any of the known laws of physics, although the expenditures of energy needed for accelerating to such velocities are enormous. A number of technological breakthroughs will be required before such flights may be considered practicable.

Once we have a capability for sending small packages over interstellar distances at velocities of the order of one-tenth the speed of light or higher, we can consider sending unmanned probes to the vicinity of the nearest promising candidate stars to report back to Earth the prevailing conditions. Favorable reports could subsequently be followed by manned expeditions to those systems showing the most promise in the light of our then greatly enhanced knowledge of astrophysics and general planetology.

The requirements for speed and ingenuity will depend on how far away the next promising candidate happens to be and on how determined people are to make the trip. If one is willing to spend 20 years *en route*, then trips to stars 4 light-years away could be made at one-fifth the speed of light (average); trips to stars 10 light-years away could be made at one-half the speed of light; trips to stars 15 light-years away could be made at three-quarters the speed of light. At velocities much higher than this, however, relativistic time contractions begin to become evident, so that for a particular round trip, less time would seem to have elapsed for a traveler than for an Earthbound observer. Thus enormous distances (from the viewpoint of the Earthbound observer) could be traversed in 20 years (from the standpoint of the traveler) if velocities very close to the speed of light are attainable.

Even at much lower vehicle velocities, however, the experienced duration of the trip could possibly be shortened for the traveler through the development of "hibernation" techniques, or their equivalent.

KINDS OF HABITABLE PLANETS

The most common kinds of habitable planets, if the ideas developed here are essentially correct, should be similar to the Earth in many respects. (The one mild peculiarity of the Earth is its rather large natural satellite.) The typical habitable planet should be of much the same mass as the Earth, although smaller on the average, and it should have a similar atmosphere, a similar night-day cycle, a sun of similar size and appearance, a mild inclination to its equator, and a moderate eccentricity to its orbit. Seasons should be part of the common experience of the inhabitants, as well as oceans of water, winds, sunsets, rainbows, beaches, blue skies, starry nights, deserts, mountains, volcanoes, earthquakes, rain, lightning, rivers, clouds, and snow and ice in the cold regions. In short, most of the physical and meteorological phenomena with which we are familiar should also be known on most other habitable planets. When it comes to the living things indigenous to the planets, of course, these might differ widely, depending on the course that evolution happened to take in each special circumstance. Even so, on each planet one would expect to find organisms carrying on photosynthesis and animal forms capable of invading practically every conceivable ecological niche: marine forms, land creatures, aerial forms, *et cetera*. In spite of differences in detail, certain basic kinds of life forms would be expected to display some common characteristics—for example, fast-swimming marine forms would be streamlined, land animals would typically have legs, and fast-moving aerial forms would have wings.

On no other planet, however, would one expect to find any of the phyla, classes, orders, families, genera, or species of plants or animals with which we are familiar on the surface of the Earth. From the smallest bacteria to the largest mammals (cetaceans), these are unique products of the Earth. Each planet on which living things have evolved must have its own peculiar classification of organisms (taxonomy). There must of necessity be autotrophs (life forms that use only inorganic nutrients), and one would also expect to find heterotrophs (parasitic life forms that use autotrophs or other heterotrophs for food).

Apart from indigenous life forms, most habitable planets should share many physical characteristics. There should also be many unusual and rarer kinds of habitable planets, however, among the more than half a billion habitable planets that have just been computed to exist in our Galaxy. Some of the special types might include those described below.

A *satellite planet* would be a habitable planet in orbit around a massive giant planet (like Jupiter), with its rotation stopped with respect to the companion but rotating with respect to its sun. Such a planet would have unusual cycles of light and dark on the side facing its companion, eclipses of the sun occurring every day (unless its orbit was quite markedly inclined to the orbit of its companion), and nights dominated by the presence of its extremely large and luminous "moon." The side away from the companion would have more normal day-night cycles.

Twin planets would be two habitable planets revolving around their common center of mass, their rotation stopped with respect to each other.

A *planet with two suns* would be a habitable planet in orbit around two stars that are very close to each other. Close binaries (separated by a few million miles, say) would produce a somewhat complicated sunrise-sunset pattern for a habitable planet in orbit around them and an interesting variation in light intensity as the stars eclipsed each other; but otherwise they would not necessarily greatly affect conditions on the surface of the planet.

A *planet in a widely separated binary system*, a habitable planet in orbit around one star in a binary system, would have very bright nights during those parts of the year when it was passing between the two stars, and "normal" dark nights only when the companion, as well as the primary, was beneath the horizon. From a habitable planet in orbit around Alpha Centauri A, for example, the companion B would range in magnitude from −18 to −20, depending on whether the two stars were at apastron or periastron. (Compare these figures to the light from the full Moon at magnitude −12 and from the Sun at magnitude −26.8.)

A *planet with very high equatorial inclination* must lie within a narrow range of orbital radius to be habitable at all, and then only a small fraction of its surface might be habitable. A planet with an equatorial inclination of 75 degrees at the optimal distance from its sun, for example, would be habitable only between latitudes 14°N and 14°S. Higher latitudes would be excessively cold during the winter season. The apparent seasonal motions of the sun and the seasonal temperature variations would be extreme.

A *planet with two habitable belts* would, most likely, be found among planets having low equatorial inclinations but orbiting near the inner edge of the ecosphere. Such planets are excessively hot near the equator, hence the habitable regions would be in the intermediate or high latitudes only. A planet having the same equatorial inclination as the Earth (23.5 degrees), but orbiting at a distance from its primary where it received 30 per cent more radiation than the Earth, would be habitable only between latitudes 51° to 66° (both north and south), with the wide belt from 51°S to 51°N being intolerably hot during much of the year. Under such circumstances, with two widely separated regions of habitability, it is quite probable that land life forms would evolve more or less independently in the two regions. Marine life and some aerial forms might migrate between the two, but land migrations would be fairly effectively stopped by a heat barrier consisting mainly of deserts, possibly with small pockets of habitability on mountain plateaus.

Habitable planets with rings are another possibility. There are many unknowns associated with the origin and composition of the beautiful ring system of Saturn. An important feature is that the rings all lie completely within Roche's limit (approximately 2.45 times the planetary radius from the center of the planet). It is quite probable that massive oblate planets are more likely to possess rings than planets in the habitable class; yet it is reasonable to suppose that some habitable planets may also have flat equatorial rings within their Roche limits, although these rings would probably not be as densely populated as the rings of Saturn.

Other special types of habitable planets may also exist. Since oceans are believed to be products of volcanic activity and this, in turn, is possibly a function of planetary mass, high-*g* and low-*g* planets in the habitable range may be more or less identical with largely oceanic planets and largely dry-land planets. Planets with more extensive oceans than the Earth—those having, say, 90 per cent ocean and 10 per cent dry land—might well have quite widely separated continents with no land bridges. Fairly independent evolutionary courses might be followed by the land life forms in such extreme isolation from other forms on the same planet. On the other hand, planets having substantially less oceanic water than the Earth might well have noninterconnecting oceans, with isolated forms of marine life following independent evolutionary paths. In the absence of worldwide oceanic circulation, there would be less moderation of temperature cycles (that is, more continental, as opposed to oceanic, climates). A high fraction of the land surface probably would be desert and the main habitable regions would tend to be close to the landlocked seas.

Our Sun is situated in a fairly sparsely populated section of the Galaxy—in one of the spiral arms fairly well out toward the edge of the disk. Our night sky, therefore, is relatively lightly sprinkled with stars in comparison with what would be seen from planets located in certain other positions in the Galaxy. Something like twenty-five hundred stars brighter than magnitude 6.5 are visible to the naked eye from a given spot on the Earth on a clear night. Much more spectacular night skies would be seen from habitable planets located in globular clusters or from habitable planets near the center of the Galaxy. In fact, Isaac Asimov (1958) has estimated that about two million stars of magnitudes brighter than 6.5 would be visible above the horizon from a habitable planet located near the galactic center. As he has pointed out, the starlight in such a sky would be roughly equal to the light of the full Moon as seen from the Earth. A correction should perhaps be made in his estimate to account for the fact that the scattered light in the sky would prevent one from seeing any but the brightest stars. Stars fainter than, perhaps, magnitude 2.5 could not be seen at all against the over-all luminosity of the night sky. The stars brighter than this limiting magnitude would number approximately thirty thousand, however, or about ten times as many as we can see on the darkest night.

On the other hand, habitable planets around stars imbedded in some of the dusty regions (dark nebulae) of the Galaxy might have almost no stars at all in their night skies; and habitable planets around stars at the very edge of the Galaxy would have stars in one half of the celestial sphere but none in the other half. The only lights in the night sky looking away from the Galaxy (apart from local planets) would come from the globular clusters that surround our Galaxy or from distant island universes, only a few of which are faintly visible to the naked eye.

CHANGES IN MEN IN NEW ENVIRONMENTS

Human beings, together with all other life forms on the Earth, are very well adapted to what we think of as a normal environment. Although the normal environments of the Eskimos, of the Australian aborigines, of the pygmies of Africa, or the Indians of the high Andes are quite different from one another, they all fall into a fairly narrow range when viewed from the standpoint of planetary parameters. All of the peoples mentioned above occupy more or less marginal ecological niches, but they have become adjusted to them gradually. This adjustment probably has come about through many generations of selection for individuals who could tolerate exceptionally well the environmental extremes of temperatures, the dietary limitations, the dryness, or the low oxygen partial pressure, as the case happened to be.

Colonizers of other planets will encounter even more varied environmental conditions than those existing on the Earth's surface. The transitions, however, unlike those made by peoples in the past history of the Earth, will of necessity be abrupt, rather than gradual. These colonizers will, of course, have the advantages of a very high level of technology to assist them in making the transitions. Even so, unless close contact is maintained with the population pool of the Earth, very profound genetic shifts will undoubtedly occur in a relatively short time if the environmental conditions are markedly different from those of the Earth.

In future times, when interstellar trips become possible, there may be circumstances under which an expedition locates a habitable planet and then, through accident or design, is cut off from communication with the rest of humanity for several hundred years.

Imagine such a colony marooned on a 1.5-*g* planet, for instance. Assuming that the colony is able to survive and multiply, there would inevitably be a premium on muscular strength, short reaction times, and accurate judgment in moving about. There would also be a premium on strong internal constitutions. Because of the high surface gravity, accidental falls would be more dangerous—more likely to be fatal or crippling.

Sprains, strained muscles, hemorrhoids, fallen internal organs, back, foot, and leg ailments, varicose veins, and certain difficulties of pregnancy would be more prevalent than in the gentler 1-*g* environment of the Earth. Thus there would be an inexorable selection pressure continually favoring those individuals best equipped to deal with the problems of a high gravitational force.

After a number of generations under these environmental conditions, what would the population look like? Those individuals best able to thrive would probably tend to have shorter arms and legs, be more compactly constructed, and have heavier bones than any population normally found on Earth. Experiments with chickens conducted by Smith and others (Wunder, 1961) have shown that animals raised under high gravity exhibit an increase in relative heart and leg sizes. These observations might also apply to human beings. Because of the constant drag of gravity, they would tend to have better muscular development and less unsupported external fatty tissue. Since it would be an advantage during pregnancy to have small babies rather than large ones, the average height of the adult population would probably tend to decrease to some optimal level.

Since objects fall much faster when surface gravity is high, selection would also tend to favor people with unusually rapid muscular reactions. With changes in appearance, the accepted standards of feminine attractiveness would tend to change also. In addition, if isolation were continued for a long enough period of time, there would inevitably be various kinds of minor genetic drifts in unpredictable directions. Such drifts have been observed statistically in small, partially isolated Earth populations, for example, certain Pennsylvania Dutch communities. If continued for a sufficient time, enough genetic changes would accumulate so that the isolated population would no longer be genetically compatible with the population of the Earth when the members again made contact. Thus human speciation could result from interstellar travel; and while this would probably require separations of many thousands of years, it would probably occur much more rapidly than it would under conditions of isolation on the Earth.

Other environmental conditions would result in quite different changes to human populations. A colony isolated on a low-gravity planet—one with a surface gravity of, say, three-quarters of Earth normal—would be exposed to a reduced stress from gravity, but it might be exposed to a reduced oxygen partial pressure. Selection processes might favor those individuals having more efficient respiratory systems and, possibly, those having larger rib cages, lungs, and respiratory tidal volumes. Under less stringent gravitational stresses, those with slender physiques would have no great genetic disadvantages, and any secular changes in the composition of the population with respect to body type would probably hinge on other factors.

On a small planet, having a thinner atmosphere and probably a weaker magnetic field than the Earth, the normal background radiation level might well be substantially higher than that at the Earth's sea level, for two reasons. First, as a result of less intense gravitational fractionation of rocky materials in the body of the planet during the formation period, the proportion of heavy minerals, including radioactive materials, in the crust might be higher. Second, with less shielding from flare protons from the primary and from galactic cosmic particles, there would be a greater influx of energetic particles from space. Accordingly, mutation rates would be expected to be higher, and possibly evolution would proceed at an accelerated rate.

Even without genetic or muscular changes, human capabilities for physical action would be markedly changed on planets having gravitational accelerations different from Earth's. This may be demonstrated with reference to track and field records and various athletic events. For events such as the shot-put, the javelin throw, and the discus throw, the maximum range of throw should be inversely proportional to the g level. If a champion athlete can throw a javelin 281 feet on the Earth, for example, he could throw it about 375 feet on a planet with a surface gravity of three-quarters Earth normal, but only 187 feet on a 1.5-g planet. For the high jump, the relationship is not as simple because a man must lift his own center of mass about a foot above the recorded height of the bar, yet his center of mass is already about 3 feet above the ground when he starts his spring into the air. If we assume that the force in the system is constant from planet to planet, that the jumper crouches $1\frac{1}{2}$ feet before making his jump, and that the world's record for the standing high jump is 5.5 feet, then the height, h, recorded for the jump would be $h=(5/g)+0.5$, for g relative to Earth normal and h in feet. Consequently, on a 0.75-g planet, the record standing high jump should be about 7.1 feet, while on a 1.5-g planet, it should be 3.8 feet.

For running events, the relationships are less clear, although it may be seen intuitively that men could run faster on low-g planets and less fast on high-g planets than they can on the Earth's surface.

CHAPTER 8

An Appreciation of the Earth

We take our home for granted most of the time. We complain about the weather, ignore the splendor of our sunsets, the scenery, and the natural beauties of the lands and seas around us, and cease to be impressed by the diversity of living species that the Earth supports. This is natural, of course, since we are all products of the Earth and have evolved in conformity with the existing environment. It is our natural habitat, and all of it seems very commonplace and normal. Yet how different our world would be if some of the astronomical parameters were changed even slightly.

Suppose that, with everything else being the same, the Earth had started out with twice its present mass, giving a surface gravity of 1.38 times Earth normal. Would the progression of animal life from sea to land have been so rapid? While the evolution of marine life would not have been greatly changed, land forms would have to be more sturdily constructed, with a lower center of mass. Trees would tend to be shorter and to have strongly buttressed trunks. Land animals would tend to develop heavier leg bones and heavier musculature. The development of flying forms would certainly have been different, to conform with the denser air (more aerodynamic drag at a given velocity) and the higher gravity (more lifting surface necessary to support a given mass). A number of opposing forces would have changed the face of the land. Mountain-forming activity might be increased, but mountains could not thrust so high and still have the structural strength to support their own weight; raindrop and stream erosion would be magnified, but the steeper density gradient in the atmosphere would change the weather patterns; wave heights in the oceans would be lower, and spray trajectories would be shortened, resulting in less evaporation and a drier atmosphere; and cloud decks would tend to be lower. The land-sea ratio would probably be smaller. The length of the sidereal month would shorten from 27.3 to 19.4 days (if the Moon's distance remained the same). There would be differences in the Earth's magnetic field, the thickness of its crust, the size of its core, the distribution of mineral deposits in the crust, the level of radioactivity in the rocks, and the size of the ice caps on islands in the polar regions. Certainly man's counterpart (assuming that such a species would have evolved in this environment) would be quite different in appearance and have quite different cultural patterns.

Conversely, suppose that the Earth had started out with half its present mass, resulting in a surface gravity of 0.73 times Earth normal. Again the course of evolution and geological history would have changed under the influences of the lower gravity, the thinner atmosphere, the reduced erosion by falling water, and the probably increased level of background radiation due to more crustal radioactivity and solar cosmic particles. Would evolution have proceeded more rapidly? Would the progression from sea to land and the entry of animal forms into the ecological niches open to airborne species have occurred earlier? Undoubtedly animal skeletons would be lighter, and trees would be generally taller and more spindly; and again, man's counterpart, evolved on such a planet, would be different in many ways.

What if the inclination of the Earth's equator initially had been 60 degrees instead of 23.5 degrees? Seasonal weather changes would then be all but intolerable, and the only climatic region suitable for life as we know it would be in a narrow belt within about 5 degrees of the equator. The rest of the planet would be either too hot or too cold during most of the year, and with such a narrow habitable range, it is probable that life would have had difficulty getting started and, once started, would have tended to evolve but slowly.

Starting out with an inclination of 0 degrees would have influenced the course of development of the Earth's life forms in only a minor way. Seasons would be an unknown phenomenon; weather would undoubtedly be far more predictable and constant from day to day. All latitudes would enjoy a constant spring. The region within 12 degrees of the equator would become too hot for habitability but, in partial compensation, some regions closer to the poles would become more habitable than they are now.

Suppose the Earth's mean distance from the Sun were 10 per cent less than it is at present. Less than 20 per cent of the surface area (that between latitudes 45 degrees and 64 degrees) would then be habitable. Thus there would be two narrow land regions favorable to life separated by a wide and intolerably hot barrier. Land life could evolve independently in these two regions. The polar ice would not be present, so the ocean level would be higher than it is now, thus decreasing the land area.

If the Earth were 10 per cent farther away from the Sun than it is, the habitable regions would be those within 47 degrees of the equator. (The present limit of habitability is assumed to be, on an average, within 60 degrees of the equator.)

If the Earth's rotation rate were increased so as to make the day 3 hours long instead of 24 hours, the oblateness would be pronounced, and changes of gravity as a function of latitude would be a common part of a traveler's experience. Day-to-night temperature differences would become small.

On the other hand, if the Earth's rotation rate were slowed to make the day 100 hours in length, day-to-night temperature changes would be extreme; weather cycles would have a more pronounced diurnal pattern. The Sun would seem to crawl across the sky, and few life forms on land could tolerate either the heat of the long day or the cold of the long night.

The effects of reducing the eccentricity of the Earth's orbit to 0 (from its present value of 0.0167) would be scarcely noticeable. If orbital eccentricity were increased to 0.2 without altering the length of the semi-major axis (making perihelion coincide with summer solstice in the Northern Hemisphere to accentuate the effects), the habitability apparently would not be affected in any significant manner.

Increasing the mass of the Sun by 20 per cent (and moving the Earth's orbit out to 1.408 astronomical units to keep the solar constant at its present level) would increase the period of revolution to 1.54 years and decrease the Sun's apparent angular diameter to 26 minutes of arc (from its present 32 minutes of arc). Our primary would then be a class F5 star with a total main-sequence lifetime of about 5.4 billion years. If the age of the solar system were 4.5 billion years, then the Earth, under these conditions, could look forward to another billion years of history. Since neither of these numbers is known to the implied accuracy, however, a 10 per cent error in each in the wrong direction could mean that the end was very near indeed. An F5 star may well be more "active" than our Sun, thus producing a higher exosphere temperature in the planetary atmosphere; but this subject is so little understood at present that no conclusions can be drawn. Presumably, apart from the longer year, the smaller apparent size of the Sun, its more pronounced whiteness, and the "imminence" of doom, life could be much the same.

If the mass of the Sun were reduced by 20 per cent (this time decreasing the Earth's orbital dimensions to compensate), the new orbital distance would be 0.654 astronomical unit. The year's length would then become 0.59 year (215 days), and the Sun's apparent angular diameter, 41 minutes of arc. The primary would be of spectral type G8 (slightly yellower than our Sun is now) with a main-sequence lifetime in excess of 20 billion years. The ocean tides due to the primary would be about equal to those due to the Moon; thus spring tides would be somewhat higher and neap tides lower than they are at present.

What if the Moon had been located much closer to the Earth than it is, say, about 95,000 miles away instead of 239,000 miles? The tidal braking force would probably have been sufficient to halt the rotation of the Earth with respect to the Moon, and the Earth's day would equal its month, now 6.9 days in length (sidereal). Consequently, the Earth would be uninhabitable.

Moving the Moon farther away than it is would have much less profound results: the month would merely be longer and the tides lower. Beyond a radius of about 446,000 miles, the Earth can not hold a satellite on a circular orbit.

Increasing the mass of the Moon by a factor of 10 at its present distance would have an effect similar to that of reducing its distance. However, the Earth's day and month would then be equal to 26 days. Decreasing the Moon's mass would affect only the tides.

What if the properties of some of the other planets of the solar system were changed? Suppose the mass of Jupiter were increased by a factor of 1050, making it essentially a replica of the Sun. The Earth could still occupy its present orbit around the Sun, but our sky would be enriched by the presence of an extremely bright star, or second sun, of magnitude −23.7, which would supply at most only 6 per cent as much heat as the Sun. Mercury and Venus could also keep their present orbits; the remaining planets could not, although those exterior to Saturn could take up new orbits around the new center of mass.

All in all, the Earth is a wonderful planet to live on, just the way it is. Almost any change in its physical properties, position, or orientation would be for the worse. We are not likely to find a planet that suits us better, although at some future time there may be men who prefer to live on other planets. At the present time, however, the Earth is the only home we have; we would do well to conserve its treasures and to use its resources intelligently.

CHAPTER 9

Human Destiny

For a long time to come, the major part of the human race will continue to live on the surface of the Earth. Yet, at some time in the future, new human colonies may be started on planets of other stars, and, eventually, the number of human beings living elsewhere may exceed the population of the Earth.

In his recent book, *The Next Million Years,* Charles G. Darwin did not even mention space flight but confined his attention to the future history of the human population of the Earth. In the long run, however, space flight may prove to be the most significant departure in the history of mankind. This may not necessarily be true for the vast majority of the Earth's population who will remain at home; but it certainly should be true for those few (and their descendants) who will journey into space to find new homes among the stars and start new colonies and new bases from which to launch still other expeditions into a gradually expanding volume of space, the next frontier. The population of the Earth can not continue to grow indefinitely at the present rate (between 1½ and 2 per cent each year). An upper limit must be reached within the next several hundred years, when the Earth's population will somehow become more or less stabilized numerically (barring a man-made catastrophe). Whether the Earth will be such a pleasant place to live at that time is problematical. It will undoubtedly be much more crowded than it is now, and there will inevitably be increasing incentives for pioneers to seek new lives among the stars for themselves and their families. However, as others have already pointed out, we can not look to space flight to solve the problem of the Earth's population explosion. This is obvious, of course, from the present rate of increase in the world's population. In mid-1962 the world's population reached the 3-billion mark, and the net annual rate of increase was estimated at 1.8 per cent, or over 50 million people per year. Just to hold the Earth's population constant at the present time would require the emigration of almost 150,000 people per day—clearly not a reasonable concept. In another century, if the present rate of natural increase continued, the emigration rate would have to be stepped up to 900,000 per day to keep the Earth's population constant at 18 billion people. Admittedly, however, experience has shown that it is dangerous to assume that present population growth trends can be used with confidence to predict future population levels.

If man learns to travel through space at, say, one-quarter or one-half the speed of light, even with long planetary stopovers on star-hopping expeditions making the net advance only one-tenth the speed of light, the entire Galaxy could be explored and all its habitable planets settled within the next million years. Unquestionably, many technological advances will occur before this much time has elapsed, and the spread of mankind throughout the Galaxy may take place more rapidly. The future history of man may well be written among the stars.

Each stage in the progress of man as he star-hops into new, unexplored regions of the Galaxy will be accompanied by an important kind of distillation process. Always, those volunteering for the next expedition into the unknown will tend to be adventurous, self-reliant, inquisitive, courageous, and hardy pioneers, while those selected to go will be chosen on the basis of good health, high professional competence, emotional stability, reliability of judgment, *et cetera*. This is based on the assumption that such expeditions will always be large, expensive projects—projects too important, whether privately or publicly financed, to be entrusted to any but the most competent and reliable people. And, in the main, these characteristics will be passed on to their descendants, so a kind of selection process will take place, with those at the frontier of the wave through the Galaxy always representing some of the best qualities of mankind.

APPENDIX

Data on the 25 Principal Bodies of the Solar System of Mass Greater Than 10^{23} Grams.

APPENDIX

Data on the 25 Principal Bodies of the Solar System (of Mass Greater Than 10^{23} Grams)

Body	Mass					Mean radius		
	grams	relative to Sun		relative to Earth	relative to primary	cm	mi	relative to Earth
		μ^{-1}	μ					
Sun	1.991×10^{33}	1.000	1.000	333,400	. . .	6.960×10^{10}	432,500	109
Planets:								
Mercury	2.765×10^{26}	7,200,000 (?)	1.39×10^{-7} (?)	0.0463 (?)	. . .	2.50×10^{8}	1,555	0.392
Venus	4.876×10^{27}	408,600	2.446×10^{-6}	0.816	. . .	6.033×10^{8}	3,750	0.946
Earth (E)	5.977×10^{27}	333,400	3.00×10^{-6}	1.000	. . .	6.371×10^{8}	3,960	1.000
Mars	6.445×10^{26}	3.09×10^{6}	3.235×10^{-7}	0.1077	. . .	3.38×10^{8}	2,100	0.530
Jupiter (J)	1.903×10^{30}	1047.4	9.55×10^{-4}	317.4	. . .	6.95×10^{9}	43,150	10.9
Saturn (S)	5.695×10^{29}	3497.6	2.86×10^{-4}	95.03	. . .	5.81×10^{9}	36,100	9.13
Uranus	8.69×10^{28}	22,934	4.36×10^{-5}	14.5	. . .	2.45×10^{9}	15.250	3.85
Neptune (N)	1.048×10^{29}	19,000	5.295×10^{-5}	17.5	. . .	2.36×10^{9}	14,660	3.70
Pluto	5×10^{27} (?)	4×10^{5} (?)	2.5×10^{-6} (?)	0.8 (?)	. . .	(?)	(?)	(?)
Satellites:								
E–1 Moon	7.35×10^{25}	. . .	. . .	0.01229	0.01229	1.738×10^{8}	1,080	0.273
J–1 Io	7.25×10^{25} (?)	. . .	. . .	1.21×10^{-2} (?)	3.81×10^{-5} (?)	1.66×10^{8} (?)	1,030 (?)	0.251 (?)
J–2 Europa	4.72×10^{25} (?)	. . .	. . .	7.9×10^{-3} (?)	2.48×10^{-5} (?)	1.44×10^{8} (?)	900 (?)	0.226 (?)
J–3 Ganymede	1.55×10^{26} (?)	. . .	. . .	2.6×10^{-2} (?)	8.17×10^{-5} (?)	2.47×10^{8} (?)	1,540 (?)	0.388 (?)
J–4 Callisto	9.68×10^{25} (?)	. . .	. . .	1.62×10^{-2} (?)	5.09×10^{-5} (?)	2.34×10^{8} (?)	1,450 (?)	0.367 (?)
S–3 Tethys	6.5×10^{23} (?)	. . .	. . .	1.09×10^{-4} (?)	1.14×10^{-6} (?)	6×10^{7} (?)	370 (?)	0.094 (?)
S–4 Dione	1.04×10^{24} (?)	. . .	. . .	1.74×10^{-4} (?)	1.82×10^{-6} (?)	6.5×10^{7} (?)	400 (?)	0.102 (?)
S–5 Rhea	2.3×10^{24} (?)	. . .	. . .	3.85×10^{-4} (?)	4.04×10^{-6} (?)	9×10^{7} (?)	560 (?)	0.141 (?)
S–6 Titan	1.37×10^{26} (?)	. . .	. . .	2.3×10^{-2} (?)	2.41×10^{-4} (?)	2.5×10^{8} (?)	1,550 (?)	0.392 (?)
S–7 Hyperion	1.1×10^{23} (?)	. . .	. . .	1.84×10^{-5} (?)	1.93×10^{-7} (?)	2×10^{7} (?)	120 (?)	0.031 (?)
S–8 Iapetus	5×10^{24} (?)	. . .	. . .	8.4×10^{-4} (?)	8.8×10^{-6} (?)	6×10^{7} (?)	370 (?)	0.094 (?)
N–1 Triton	1.38×10^{26} (?)	. . .	. . .	2.31×10^{-2} (?)	1.32×10^{-3} (?)	2×10^{8} (?)	1,240 (?)	0.31 (?)
Asteroids:								
Ceres	8×10^{23} (?)	. . .	. . .	1.34×10^{-4} (?)	. . .	3.7×10^{7} (?)	230 (?)	0.058 (?)
Pallas	2×10^{23} (?)	. . .	. . .	3.35×10^{-5} (?)	. . .	2.4×10^{7} (?)	150 (?)	0.038 (?)
Vesta	1×10^{23} (?)	. . .	. . .	1.67×10^{-5} (?)	. . .	1.9×10^{7} (?)	60 (?)	0.030 (?)

Body	Rotation			Semimajor axis of orbit (distance to primary)			Distance to primary	
	length of sidereal day	angular velocity at equator, ω (rad/sec)	ω^2 (rad²/sec²)	cm	A.U.	mi	minimum (cm)	maximum (cm)
Sun	25.1 days	2.90×10^{-6}	8.41×10^{-12}	...	...	...	...	...
Planets:								
Mercury	87.97 days	8.28×10^{-7}	6.83×10^{-13}	5.79×10^{12}	0.387	3.59×10^{7}	4.60×10^{12}	6.98×10^{12}
Venus	(?)	(?)	(?)	1.082×10^{13}	0.723	6.71×10^{7}	1.075×10^{13}	1.09×10^{13}
Earth (E)	23.935 hours	7.29×10^{-5}	5.325×10^{-9}	1.495×10^{13}	1.000	9.29×10^{7}	1.471×10^{13}	1.521×10^{13}
Mars	24.623 hours	7.1×10^{-5}	5.03×10^{-9}	2.279×10^{13}	1.524	1.416×10^{8}	2.065×10^{13}	2.49×10^{13}
Jupiter (J)	9.842 hours	1.775×10^{-4}	3.148×10^{-8}	7.783×10^{13}	5.203	4.83×10^{8}	7.40×10^{13}	8.15×10^{13}
Saturn (S)	10.23 hours	1.707×10^{-4}	2.91×10^{-8}	1.428×10^{14}	9.546	8.86×10^{8}	1.347×10^{14}	1.507×10^{14}
Uranus	10.82 hours	1.604×10^{-4}	2.58×10^{-8}	2.872×10^{14}	19.20	1.783×10^{9}	2.735×10^{14}	3.005×10^{14}
Neptune (N)	{15.67 hours 12.43 hours	1.114×10^{-4} 1.403×10^{-4}	1.242×10^{-8} 1.97×10^{-8}}	4.498×10^{14}	30.09	2.795×10^{9}	4.46×10^{14}	4.54×10^{14}
Pluto	16 hours (?)	...	...	5.910×10^{14}	39.5	3.67×10^{9}	4.45×10^{14}	7.38×10^{14}
Satellites:								
E–1 Moon	27.32 days	2.66×10^{-6}	7.09×10^{-12}	3.844×10^{10}	2.571×10^{-3}	2.385×10^{5}	3.63×10^{10}	4.06×10^{10}
J–1 Io	...	...	...	4.218×10^{10}	2.82×10^{-3}	2.62×10^{5}	4.215×10^{10}	4.22×10^{10}
J–2 Europa	...	...	...	6.714×10^{10}	4.49×10^{-3}	4.17×10^{5}	6.66×10^{10}	6.76×10^{10}
J–3 Ganymede	...	...	...	1.071×10^{11}	7.16×10^{-3}	6.65×10^{5}	9.85×10^{10}	1.157×10^{11}
J–4 Callisto	...	...	...	1.884×10^{11}	1.259×10^{-2}	1.17×10^{6}	...	...
S–3 Tethys	...	...	...	2.948×10^{10}	1.97×10^{-3}	1.83×10^{5}	2.948×10^{10}	2.948×10^{10}
S–4 Dione	...	...	...	3.777×10^{10}	2.52×10^{-3}	2.34×10^{5}	3.77×10^{10}	3.785×10^{10}
S–5 Rhea	...	...	...	5.275×10^{10}	3.52×10^{-3}	3.27×10^{5}	5.27×10^{10}	5.28×10^{10}
S–6 Titan	...	...	...	1.223×10^{11}	8.17×10^{-3}	7.59×10^{5}	1.19×10^{11}	1.26×10^{11}
S–7 Hyperion	...	...	...	1.484×10^{11}	9.89×10^{-3}	9.19×10^{5}	1.32×10^{11}	1.647×10^{11}
S–8 Iapetus	...	...	...	3.563×10^{11}	2.38×10^{-2}	2.21×10^{6}	3.555×10^{11}	3.67×10^{11}
N–1 Triton	...	...	...	3.53×10^{10}	2.36×10^{-3}	2.19×10^{5}	3.53×10^{10}	3.53×10^{10}
Asteroids:								
Ceres	9.08 hours	...	...	4.14×10^{13}	2.767	2.57×10^{8}	3.81×10^{13}	4.46×10^{13}
Pallas	...	...	...	4.14×10^{13}	2.767	2.57×10^{8}	3.17×10^{13}	5.14×10^{13}
Vesta	...	...	...	3.53×10^{13}	2.361	2.19×10^{8}	3.31×10^{13}	3.84×10^{13}

Body	Polar radius (cm)	Equatorial radius (cm)	Equatorial angular diameter at mean C or O with Earth (sec)	Volume		Mean density (g/cm^3)	Oblateness, ε
				cm^3	relative to Earth		
Sun	6.96×10^{10}	6.96×10^{10}	1919	1.412×10^{33}	1.303×10^{6}	1.41	0.0
Planets:							
Mercury	2.50×10^{8} (?)	2.50×10^{8} (?)	10.90	6.55×10^{25} (?)	0.060 (?)	4.24 (?)	0.0
Venus	6.03×10^{8} (?)	6.03×10^{8} (?)	61.4	9.18×10^{26}	0.85	5.32	0.0
Earth (E)	6.357×10^{8}	6.378×10^{8}	...	1.083×10^{27}	1.000	5.52	0.00336
Mars	3.36×10^{8}	3.38×10^{8}	17.88	1.61×10^{26}	0.149	4.0	0.0052
Jupiter (J)	6.67×10^{9}	7.12×10^{9}	46.86	1.42×10^{30}	1308	1.34	0.062
Saturn (S)	5.43×10^{9}	6.01×10^{9}	19.52	8.23×10^{29}	761	0.69	0.096
Uranus	2.46×10^{9}	2.59×10^{9}	3.62	6.88×10^{28}	63.5	1.26	0.05
	2.29×10^{9}	2.41×10^{9}		5.55×10^{28}	51.3	1.56	0.072
Neptune (N)	2.44×10^{9}	2.52×10^{9}	2.12	6.5×10^{28}	60	1.61	0.02
	2.17×10^{9}	2.25×10^{9}		4.62×10^{28}	42.6	2.27	0.0333
Pluto	(?)	7.2×10^{8} (?)	0.52 (?)	(?)	(?)	(?)	(?)
Satellites:							
E–1 Moon	1.738×10^{8}	1.738×10^{8}	1866	2.199×10^{25}	0.0203	3.34	0.0
J–1 Io	...	...	...	...	...	...	...
J–2 Europa	...	...	...	...	...	...	...
J–3 Ganymede	...	...	...	...	...	...	...
J–4 Callisto	...	...	...	...	...	...	...
S–3 Tethys	...	...	...	...	...	...	...
S–4 Dione	...	...	...	...	...	...	...
S–5 Rhea	...	...	...	...	...	...	...
S–6 Titan	...	...	...	...	...	...	...
S–7 Hyperion	...	...	...	...	...	...	...
S–8 Iapetus	...	...	...	...	...	...	...
N–1 Triton	...	...	...	...	...	...	...
Asteroids:							
Ceres	...	...	...	2.2×10^{23}(?)	...	...	...
Pallas	...	...	...	6×10^{22}(?)	...	...	...
Vesta	...	...	...	3×10^{22}(?)	...	...	...

Body	Mean irradiance from Sun			Surface gravity		Eccentricity of orbit, e	Inclination of equator, i (degrees)	Rotational energy (ergs)
	erg/cm² sec	cal/cm² min	relative to Earth	relative to Earth (g)	cm/sec²			
Sun	6.35×10^{10}	...	...	27.95	2.74×10^{4}	...	7.25	2.1×10^{42}
Planets:								
Mercury	9.18×10^{6}	13.16	6.68	0.30 (?)	295 (?)	0.2056	(?)	2.25×10^{30}
Venus	2.64×10^{6}	3.78	1.92	0.91	894	0.00682	(?)	(?)
Earth (E)	1.374×10^{6}	1.97	1.000	1.000	980.7	0.01675	23.45	2.155×10^{36}
Mars	5.91×10^{5}	0.847	0.43	0.38	376	0.09331	25.2	6.64×10^{34}
Jupiter (J)	5.09×10^{4}	0.073	0.037	2.68	2630	0.04833	3.12	3.72×10^{41}
Saturn (S)	1.51×10^{4}	0.0207	0.011	1.15	1130	0.05589	26.75	7.06×10^{40}
Uranus	3.71×10^{3}	0.0053	0.0027	0.99	970	0.0470	97.98	$\left\{\begin{matrix}1.78\\1.54\end{matrix}\right. \times 10^{39}$
Neptune (N)	1.51×10^{3}	0.00207	0.0011	1.28	1260	0.0087	29	$\left\{\begin{matrix}1.58\\0.80\end{matrix}\right. \times 10^{39}$
Pluto	8.8×10^{2}	0.00126	0.00064	(?)	(?)	0.247	(?)	(?)
Satellites:								
E–1 Moon	1.374×10^{6}	1.97	1.000	0.165	162	0.055	1.53	3.12×10^{30}
J–1 Io	...	...	0.037	...	...	0.0006	...	...
J–2 Europa	...	...	0.037	...	...	0.0075	...	...
J–3 Ganymede	...	...	0.037	...	...	0.0796	...	...
J–4 Callisto	...	...	0.037	...	...	...	...	...
S–3 Tethys	...	...	0.011	...	...	0.0000	...	...
S–4 Dione	...	...	0.011	...	...	0.0021	...	...
S–5 Rhea	...	...	0.011	...	...	0.0009	...	...
S–6 Titan	...	...	0.011	...	...	0.0289	...	...
S–7 Hyperion	...	...	0.011	...	...	0.110	...	...
S–8 Iapetus	...	...	0.011	...	...	0.029	...	...
N–1 Triton	...	...	0.0011	...	...	0.000	...	...
Asteroids:								
Ceres	1.8×10^{5}	0.254	0.13	...	...	0.079	...	...
Pallas	1.8×10^{5}	0.254	0.13	...	...	0.235	...	...
Vesta	...	...	...	...	...	0.088	...	...

Body	Rotational energy per unit mass (cm^2/sec^2)	Sidereal period: sec	Sidereal period: Earth (years/days)	Mean orbital motion (rad/sec)	Principal moment of inertia, C ($g\ cm^2$)	$\left(\frac{C}{Ma^2}\right)$ k_2	Visual albedo	Surface velocity of escape: (km/sec)	Surface velocity of escape: (mi/sec)	Rotational angular momentum (cm^2 g/sec)	$\frac{\omega^2}{2\pi G\bar{\rho}}$	Revolutionary angular momentum (cm^2 g/sec)
Sun	1.06×10^9	...	...	...	5×10^{53}	...	...	618	384	1.45×10^{48}	1.42×10^{-5}	...
Planets:												
Mercury	8.13×10^3	7.6×10^6	0.2408 years	8.27×10^{-7}	6.56×10^{42}	0.38 (?)	0.060	3.8 (?)	2.4 (?)	5.44×10^{36}	3.84×10^{-7}	...
Venus	(?)	1.943×10^7	0.6152 years	3.23×10^{-7}	...	...	0.61	10.4	6.45	(?)	...	1.84×10^{47}
Earth (E)	3.605×10^8	3.155×10^7	1.0000 year	1.99×10^{-7}	8.11×10^{44}	0.3340	0.36	11.2	6.95	5.91×10^{40}	0.00230	2.66×10^{47}
Mars	1.03×10^8	5.94×10^7	1.8809 years	1.06×10^{-7}	2.64×10^{43}	0.359	0.150	5.0	3.1	1.87×10^{39}	0.00300	...
Jupiter (J)	1.96×10^{11}	3.74×10^8	11.862 years	1.68×10^{-8}	2.36×10^{49}	0.241	0.41	60.5	37.6	4.19×10^{45}	0.056	1.934×10^{50}
Saturn (S)	1.24×10^{11}	9.3×10^8	29.458 years	6.75×10^{-9}	4.85×10^{48}	0.235	0.42	36	22	8.28×10^{44}	0.101	7.835×10^{49}
Uranus	$\begin{Bmatrix}2.05\\1.77\end{Bmatrix} \times 10^{10}$	2.65×10^9	84.018 years	2.37×10^{-9}	$\begin{Bmatrix}1.38\\1.19\end{Bmatrix} \times 10^{47}$	0.236	0.45	22	13.5	$\begin{Bmatrix}2.21\\1.91\end{Bmatrix} \times 10^{43}$	$\begin{Bmatrix}0.049\\0.039\end{Bmatrix}$	1.695×10^{49}
Neptune (N)	$\begin{Bmatrix}1.51\\0.76\end{Bmatrix} \times 10^{10}$	5.2×10^9	164.78 years	1.21×10^{-9}	$\begin{Bmatrix}1.60\\1.28\end{Bmatrix} \times 10^{47}$	0.241	0.54	24	15	$\begin{Bmatrix}2.25\\1.43\end{Bmatrix} \times 10^{43}$	$\begin{Bmatrix}0.0184\\0.0130\end{Bmatrix}$ $\begin{Bmatrix}0.0292\\0.0207\end{Bmatrix}$	2.565×10^{49}
Pluto	(?)	7.84×10^9	248.4 years	8.02×10^{-10}	(?)	(?)	0.16 (?)	(?)	(?)	(?)	...	...
Satellites:												
E–1 Moon	4.25×10^4	2.36×10^6	27.32 days	2.66×10^{-6}	2.64×10^{43}	0.397	0.070	2.37	1.48	2.34×10^{36}	5.06×10^{-6}	...
J–1 Io	...	1.53×10^5	1.77 days	4.11×10^{-5}	...	...	0.37	...	...	...	...	...
J–2 Europa	...	3.07×10^5	3.55 days	2.05×10^{-5}	...	...	0.39	...	...	...	...	...
J–3 Ganymede	...	6.18×10^5	7.15 days	1.02×10^{-5}	...	...	0.20	...	...	...	...	...
J–4 Callisto	...	1.44×10^6	16.69 days	4.36×10^{-6}	...	...	0.029	...	...	...	...	...
S–3 Tethys	...	1.63×10^5	1.89 days	3.86×10^{-5}	...	...	...	...	...	...	...	...
S–4 Dione	...	2.37×10^5	2.74 days	2.65×10^{-5}	...	...	...	...	...	...	...	...
S–5 Rhea	...	3.91×10^5	4.52 days	1.61×10^{-5}	...	...	...	...	...	...	...	...
S–6 Titan	...	1.38×10^6	15.95 days	4.55×10^{-6}	...	...	...	...	...	...	...	...
S–7 Hyperion	...	1.84×10^6	21.3 days	3.42×10^{-6}	...	...	...	...	...	...	...	...
S–8 Iapetus	...	6.85×10^6	79.3 days	9.19×10^{-7}	...	...	...	...	...	...	...	...
N–1 Triton	...	5.08×10^5	5.88 days	1.24×10^{-5}	...	...	...	...	...	...	...	...
Asteroids:												
Ceres	...	1.45×10^8	4.60 years	4.33×10^{-8}	...	...	0.028	...	...	...	...	...
Pallas	...	1.46×10^8	4.61 years	...	...	...	...	...	...	...	...	...
Vesta	...	1.15×10^8	3.63 years	...	...	...	...	...	...	...	...	...
											Total	3.15×10^{50}

GLOSSARY

Accretion. The process of growth by the external addition of new matter and the coherence of separate particles.

Albedo. The ratio that the light reflected from an unpolished surface bears to the total light falling upon it.

Angular momentum. Momentum of rotation; equals the product of the moment of inertia about the axis of rotation and the angular velocity. Angular momentum is a conserved quantity; it remains constant in any isolated system.

Apastron. That point in the orbit of a double star where one of the stars is farthest from its primary; opposed to periastron.

Aphelion. The point of a planet's orbit most distant from the Sun; opposed to perihelion.

Asteroid. One of the numerous small planets nearly all of whose orbits lie between those of Mars and Jupiter.

Astronomical Unit. A unit of length equal to the mean radius of the Earth's orbit; abbreviated A.U. One astronomical unit equals about 92.9 million miles, or 149.5 million kilometers.

Autotroph. A living thing using only inorganic materials as food, as opposed to heterotrophs, allotrophs, parasites, or saprophytes, which depend on other organisms for nutrition.

Binary star system. Two stars relatively close together and revolving about their common center of gravity. The stars revolve in elliptical orbits with periods ranging from a few hours to thousands of years.

Black body. An ideal, or imaginary, body that is absolutely black when cold but is a perfect absorber of radiation and, at the same time, a perfect radiator.

Bode's law (after J. E. Bode, German astronomer). An approximate empirical expression for the relative distance of planets from the Sun. This law was originated by J. D. Titius, German mathematician (1729–1796).

Capture radius. The radius that a body would need in the absence of gravitational forces to capture a particle hitting tangentially; a function of the body's mass and radius and the particle's velocity.

Celestial mechanics. A branch of astronomy dealing with the motions of celestial bodies under the forces of gravitation, for example, the orbits of planets.

Cosmogony. That part of astronomy which treats of the origin and development of the universe and its members.

Direct motion (as applied to bodies moving on orbits). In the direction of the general planetary motion. For example, looking down on the solar system from the north, the planets move about the Sun counterclockwise. The Moon moves about the Earth counterclockwise; therefore its motion is described as direct.

Eccentricity (as applied to an elliptical orbit). The ratio of the distance between the center and either focus to the semimajor axis. The eccentricity of a circle is zero; the eccentricity of a parabola, the limiting case of an ellipse, is one.

Ecliptic. The plane of the Earth's orbit.

Ecology. That branch of biology that deals with the mutual relations among organisms and between them and their environment.

Ecosphere. As used here, a region in space in the vicinity of a star in which suitable planets can have surface conditions compatible with the origin, evolution to complex forms, and continuous existence of land life, and surface conditions suitable for human beings and the ecological complex on which they depend.

Equinox. The time at which the Sun crosses the celestial equator; then the day and the night are of equal length.

Escape velocity. The speed that an object must acquire to escape from a planet's gravitation; equals $(2GM/R)^{\frac{1}{2}}$, where G is the constant of gravitation, 6.67×10^{-8} cm^3/sec^2 g; M is the planet's mass in grams; and R is the planet's radius in centimeters.

Exobiology. The study of extraterrestrial life forms; the design of experiments directed toward attempts to discover forms of life that originated elsewhere in the universe.

Exosphere. The outermost layer of a planet's atmosphere from which gases could escape to space if their molecular velocities were sufficiently high.

Flares (specifically, solar flares). Very bright spotlike outbursts on the Sun, generally observed over or near large sunspots. Flares occur at unpredictable intervals, last from a few minutes to an hour or more, and emit high-energy protons which constitute one of the more serious hazards of manned space flight.

Galaxy. A large gravitational system of stars. Hundreds of thousands of galaxies have been photographed. *Our* Galaxy, the Milky Way, of which our Sun is a member, includes all the stars that can be seen by observers on Earth without the aid of a telescope and most of the objects that can be seen through small telescopes. Our Galaxy is a spiral galaxy, having the general shape of a lens. It is about 100,000 light-years in diameter and 10,000 light-years thick. The Sun is about two-thirds the distance, or 30,000 light-years, from its center. Our Galaxy contains on the order of 100 billion stars.

General planetology. A branch of astronomy that deals with the study and interpretation of the physical and chemical properties of planets.

Geocentric. Relating to the Earth as a center or central point of reference.

Globular cluster. A group of stars clustered into a spherical or slightly flattened spheroidal shape, generally containing thousands of individual stars. The globular clusters, of which about 100 are known, are distributed spherically around the center of our Galaxy.

Gravitation. The universal attraction exerted by every particle of matter on every other particle. For two particles of matter m_1 and m_2, separated by a distance r, the force of gravitation is proportional to $(m_1 \times m_2)/r^2$.

Gravitational constant, G. Equals 6.67×10^{-8} cm^3/sec^2 g $= 6.67 \times 10^{-8}$ dynes cm^2/g^2.

Gravity. The effect, on the surface of a celestial body, of its gravitation and of the centrifugal force produced by its rotation. On the Earth's surface the value of gravity (symbol g) is about 981 cm/sec^2 or 32.17 feet/sec^2. In general, the gravity on a spherical body is equal to GM/R^2, where G is the universal constant of gravitation, M is the mass of the planet, and R is its radius.

G-suit. A tightly fitting garment designed for use by pilots of high-performance aircraft and the like to prevent the pooling of blood in the lower part of the body and thus to forestall blackout during high-acceleration pull-out maneuvers.

Heliocentric. Relating to the Sun as a center or as a central point of reference.

Heterotroph. An organism that obtains nourishment from outside sources, using for food combined organic substances, for example, animals and certain plants that partly or wholly dispense with photosynthesis.

Hypoxia. Oxygen deficiency in the blood, cells, or tissues.

Illuminance (illumination). The density of luminous flux on a surface; time rate of flow of visible light per unit of surface area.

Inclination. The angle between two planes, such as the angle between the plane of a planet's equator and the plane of its orbit or the angle between the plane of a planet's orbit and a reference plane.

Lambert body. A body with a surface that reflects light perfectly diffusely, having a dull, as opposed to a shiny (specular), surface.

Latitude. Geocentric latitude is the angle between the equatorial plane and a line from the center of the Earth passing through the place; geodetic latitude is the angle between the equatorial plane and the local vertical. Geocentric latitude and geodetic latitude are the same on a sphere, but different on an oblate spheroid.

Light-year. The distance over which light can travel in a year's time, used as a unit in expressing stellar distances. One light-year $= 0.306$ parsec $= 6.33 \times 10^4$ A.U. $= 5.88 \times 10^{12}$ miles $= 9.46 \times 10^{12}$ km $= 9.46 \times 10^{17}$ cm.

Luminosity. As applied to stars, the luminosity is the ratio of the amount of light that would reach us from a star to the amount that would reach us from the Sun if both the star and the Sun were at the same distance from us.

Magnitude. The measure of the relative brightness of a star. The absolute magnitude is the magnitude of a star as it would appear if viewed from the standard distance of 10 parsecs (32.6 light-years). The apparent magnitude is its brightness as we see it. The absolute magnitude of the Sun is $+4.8$; its apparent magnitude is -26.8. The absolute magnitude M of a star is related to its luminosity relative to the Sun L by the expression, $M = 4.8 - 2.5 \log L$. The absolute and apparent magnitudes are related to distance in parsecs D by the equation, $M = m + 5 - 5 \log D$, where m is the apparent magnitude.

Main-sequence stars. The stars that are in the smooth curve called main sequence of the Hertzsprung–Russell diagram of absolute magnitude versus spectral class; these stars are believed to be in the stable phase of their lifetimes. (After they have consumed a certain fraction of their nuclear fuel, they become unstable and go into later evolutionary stages. Their temperatures, diameters, and internal processes change rapidly, and they become red giants and variables. Finally when their nuclear fuels are exhausted, they become white dwarfs, possibly after having explosively ejected part of their mass.)

Mass ratio. In a system containing two massive bodies, the mass ratio is the ratio of the mass of the smaller to the sum of the masses of the two.

Mean free path. The average distance that a particle (for example, a molecule) travels between successive collisions with other particles of an ensemble.

Meteorite. A stony or metallic body that has fallen to the Earth from outer space.

Millibar. A unit of pressure used in meteorology, one one-thousandth of a bar; abbreviated mb. A bar is 10^6 dynes/cm^2. Therefore 1 millibar is 10^3 dynes/cm^2. Normal atmospheric pressure at the Earth's surface is 1013 millibars.

Moment of inertia. A measure of the effectiveness of mass in rotation. The moment of inertia of a homogeneous spheroid of revolution around its polar axis is $0.4\ mr^2$, where m is the body's mass and r is the equatorial radius.

Nebulae. Vast cloudy aggregations of tenuous matter in our Galaxy. Bright nebulae are made luminous by stars in their vicinity; dark nebulae are recognized by their obscuring effect on the stars behind them.

Oblateness. State of being flattened or depressed at the poles; the flattening of a spheroid. Numerically it is the difference between the equatorial and polar diameters divided by the equatorial diameter.

Orbit. The path described by a celestial body in its revolution about another under gravitational attraction.

Orbital velocity. The velocity with which a body moves in an orbit. Circular orbital velocity of a particle equals $(GM/R)^{\frac{1}{2}}$, where G is the constant of gravitation, M is the mass of the gravitating body, and R is the distance of the particle from the center of mass.

Parallax. Generally, the apparent difference in the position of a celestial body when viewed from different positions. *Heliocentric parallax* is the angle subtended by the radius of the Earth's orbit as seen from a specified star, usually measured in thousandths of a second of arc.

Parsec. A unit of measure for interstellar distances, the distance at which the heliocentric *par*allax is one *sec*ond of arc. One parsec $= 2.06 \times 10^5$ A.U. $= 3.26$ light-years $= 1.92 \times 10^{13}$ miles $= 3.08 \times 10^{13}$ km $= 3.08 \times 10^{18}$ cm.

Partial pressure. The pressure exerted by one component of a gaseous mixture.

Partial pressure equals fractional concentration of the component (by volume) times total pressure.

Perihelion. The point of a planet's orbit closest to the Sun.

Period. The time in which a planet or satellite makes a full revolution about its primary.

Photosynthesis. The formation of organic chemical compounds from water and the carbon dioxide of the air in the chlorophyll-containing tissues of plants exposed to light.

Retrograde motion. In the direction opposite to the general planetary motion.

Retrolental fibroplasia. The formation of fibrous tissue behind the lens of the eye.

Revolution. The term generally reserved for orbital motion as opposed to rotation about an axis (for example, the revolution of the Earth about the Sun).

Roche's limit. The distance from the center of a planet equal to about 2.45 times its radius, within which a liquid satellite of the same density would be broken apart by the tide-raising forces of the planet.

Roentgen. The unit used in radiology to measure the quantity of absorbed radiation.

Root-mean-square (rms) velocity. The square root of the mean of the squares of the speeds of the particles composing a system.

Rotation. The turning of a body about an axis passed through itself.

Rotational energy. If a mass whose moment of inertia about an axis is I rotates with angular velocity ω about this axis, the kinetic energy of rotation is $\frac{1}{2}I\omega^2$. For a homogeneous spheroid, the rotational energy is $0.2\ mr^2\omega^2$.

Semimajor axis. One-half the longest dimension of an ellipse.

Sidereal month. The period of a complete revolution of the Moon from the time at which its image coincides with that of any star to the time of the next such coincidence.

Solstice. One of the two moments in a year when the Sun in its apparent motion attains its maximum distance from the celestial equator.

Speciation. The evolutionary process by which species are formed; the process by which species variations become fixed.

Spectral classes. Classification of stars based mainly on a progressive change in prominence of certain properties such as color, temperature, and presence and intensity of predominance of certain spectral lines. The principal classes, in descending order of temperature and excitation, are O, B, A, F, G, K, and M. Class O stars are blue-white and very hot; the stars in Class B are also blue-white but are less hot and are sometimes referred to as helium stars (for the dominant lines in their spectra). Class A contains white stars known as hydrogen stars. Class F stars are yellow-white. Class G stars are yellow (our Sun is a member of this class). Class K stars are orange and Class M stars are red. Each class is divided into ten spectral types, each designated by a number from 0 to 9 appended to the capital letter denoting the class.

Spectroscopic double. A binary star whose components are too close to be resolved visually but are detected by the mutual shift of their spectral lines owing to their varying velocity in the line of sight.

Synodic. Pertaining to conjunction, especially to the period between two successive conjunctions of the same bodies, as of the Moon or a planet with the Sun.

Three-body problem. The problem of predicting positions and motions of each of three bodies attracting the others in accordance with the law of gravitation. No *general* mathematical solutions are available.

Universe. All of creation; everything that exists; the entire celestial cosmos.

Velocity of light. Approximately 186,000 miles per second, or 300,000 kilometers per second.

Visual double. A binary star that can be separated into two individual stars through use of the telescope.

White dwarf. A class of small, very dense white stars of low luminosity, believed to be composed of collapsed degenerate matter and to represent the final stages in the process of stellar evolution when all the nuclear fuel has been used up. White dwarfs have been called "dying stars" that are cooling off and shine only by virtue of the heat generated in their final gravitational collapse.

BIBLIOGRAPHY

Air Force Pamphlet No. AFP 160-6-1, *Threshold Limit Values for Toxic Chemicals.* U.S. Department of the Air Force, February 16, 1959.

ALLEN, C. W. *Astrophysical Quantities.* London: The Athlone Press, 1955.

AREND, S. *Com. l'Ob. R. Belg.*, No. 20 (1950).

ASIMOV, ISAAC. "Our Lonely Planet," *Astounding Science Fiction*, **62,** No. 3 (November, 1958), pp. 127–137.

BEADLE, G. W. "The Place of Genetics in Modern Biology," *Eng. Sci.*, **23,** No. 6 (March, 1960), pp. 11–17.

BERGER, R. "The Proton Irradiance of Methane, Ammonia and Water at 77° K," *Proc. Nat. Acad. Sci. U. S.*, **47,** No. 9 (September, 1961), p. 1434.

BLANCO, V. M., and MCCUSKEY, S. W. *Basic Physics of the Solar System.* Reading: Addison-Wesley Publishing Co., Inc., 1961.

BOSS, B. *General Catalog* of *33342 Stars for the Epoch 1950.* Washington: Carnegie Institution of Washington, 1936.

BULLARD, EDWARD. "The Interior of the Earth," in *The Earth as a Planet*, ed. G. P. KUIPER. Chicago: University of Chicago Press, 1954, pp. 57–137.

BUSSARD, R. W. "Galactic Matter and Interstellar Flight," *Astronaut. Acta*, **6,** No. 4 (1960), pp. 179–194.

CALVIN, M. *Chemical Evolution and the Origin of Life*, University of California Radiation Laboratory Report No. UCRL-2124 rev., August 11, 1955.

———. *Origin of Life on Earth and Elsewhere*, University of California Radiation Laboratory Report No. UCRL-9005, December, 1959.

———. "The Chemistry of Life," *Chem. Eng. News*, **39,** No. 21 (May 22, 1961), pp. 96–104.

CHAMBERLAIN, J. W. "Upper Atmospheres of the Planets," *Astrophys. J.*, **136,** No. 2 (September, 1962), pp. 582–593.

CIRA-1961 (COSPAR International Reference Atmosphere, 1961). New York: Interscience Publishers, Inc., 1961.

COCCONI, G., and MORRISON, P. "Searching for Interstellar Communications," *Nature*, **184,** No. 4690 (September 19, 1959), pp. 844–846.

CODE, C. F., WOOD, E. H., and LAMBERT, E. H. "The Limiting Effect of Centripetal Acceleration on Man's Ability To Move," *J. Aeron. Sci.*, **14,** No. 2 (February, 1947), pp. 117–123.

CULLEN, S. C., and GROSS, E. G. "The Anesthetic Properties of Xenon in Animals and Human Beings, with Additional Observations on Krypton," *Science*, **113,** No. 2942 (May 18, 1951), pp. 580–582.

DARWIN, C. G. *The Next Million Years.* Garden City: Doubleday & Company, Inc., 1952.

DAVIS, M. H. "Properties of the Martian Atmosphere," *Quarterly Technical Progress Report (4)*, RM-2816-JPL, The RAND Corporation, June 30, 1961.

DE MARCUS, W. C. "Planetary Interiors," *Encyclopedia of Physics*, **52,** ed. S. FLÜGGE. Berlin: Springer-Verlag, 1958.

DOLE, S. H. "Limits for Stable Near-circular Planetary or Satellite Orbits in the Restricted Three-body Problem," *ARS J.*, **31**, No. 2 (February, 1961), pp. 214–219.

EDSALL, J. T., and WYMAN, J. *Biophysical Chemistry*. New York: Academic Press, Inc., 1958.

FAIRHALL, L. T. *Industrial Toxicology*. Baltimore: The Williams & Wilkins Co., 1949.

FIRSOFF, V. A. *Our Neighbour Worlds*. London: Hutchinsons and Co., Ltd., 1954.

FRUTON, J. S., and SIMMONDS, S. *General Biochemistry* (2nd ed.). New York: John Wiley & Sons, Inc., 1959.

HALDANE, J. B. S. "The Origin of Life," *Rationalist Ann.* (1928), pp. 148–153.

———. "The Origin of Life," *New Biol.*, No. 16 (1954), pp. 12–27.

HENDERSON, L. J. *The Fitness of the Environment*. Boston: Beacon Press, 1958.

HERTZSPRUNG, E. "Über die Verwendung photographischer effektiver Wellenlängen zur Bestimmung von Farbenequivalenten" (Regarding the Use of Photographic Effective Wave Lengths in the Determination of Color Equivalents), *Publ. Astrophys. Obs. Potsdam*, **22** (1911).

HOPKINSON, R. G. "Glare Discomfort and Pupil Diameter," *J. Opt. Soc. Am.*, **46**, No. 8 (August, 1956), pp. 649–656.

HOROWITZ, N. H. "The Origin of Life," *Eng. Sci.*, **20**, No. 2 (November, 1956), pp. 21–25.

———. "The Origin of Life," in *Frontiers of Science*, ed. E. HUTCHINGS, JR. New York: Basic Books, Inc., 1958.

HOYLE, F. *The Nature of the Universe*. New York: Harper and Brothers, 1950.

———. *Frontiers of Astronomy*. New York: Harper and Brothers, 1955.

HUANG, S. "Occurrence of Life in the Universe," *Am. Scientist*, **47**, No. 3 (September, 1959), pp. 397–402.

INGALLS, T. H. "The Strange Case of the Blind Babies," *Sci. Am.*, **193**, No. 6 (December, 1955), pp. 40–44.

JASCHEK, C., and JASCHEK, M. "The Frequency of Spectroscopic Binaries," *Publ. Astron. Soc. Pacific*, **69**, No. 411 (December, 1957), pp. 546–551.

JEANS, J. H. *The Dynamical Theory of Gases*. London: Cambridge University Press, 1916.

———. *Astronomy and Cosmogony*. London: Cambridge University Press, 1929.

JONES, H. SPENCER. *Life on Other Worlds*. New York: New American Library of World Literature, Inc., 1949.

JONES, J. E. *Trans. Cambridge Phil. Soc.*, **22** (1923), p. 535.

JOYCE, J. *Ulysses*. New York: Random House, Inc., 1946, p. 655.

KOPAL, Z., and FIELDER, G. *Studies in Lunar Topography: I. Determination of the Heights of Mountains on the Moon*, Air Force Cambridge Research Center Report No. AFCRC-TN-59-411, ASTIA No. 217039, March, 1959.

KRAUSS, R. W., and OSRETKAR, A. "Minimum and Maximum Tolerance of Algae to Temperature and Light Intensity," in *Medical and Biological Aspects of the Energies of Space*, ed. PAUL A. CAMPBELL. New York: Columbia University Press, 1961.

KUIPER, G. P. "On the Origin of the Solar System," in *Astrophysics*, ed. J. A. HYNEK. New York: McGraw-Hill Book Co., Inc., 1951.

LAMB, H. *Higher Mechanics* (2nd ed.). London: Cambridge University Press, 1929.

LONDON, J. *A Study of the Atmospheric Heat Balance*, Report No. AFCRC-TR-57-287, Department of Meteorology and Oceanography, New York University, Air Force Cambridge Research Center, 1957.

LOWELL, P. *Mars as the Abode of Life*. New York: The Macmillan Co., 1908.

MACDONALD, G. A. "Volcanology," *Science*, **133**, No. 3454 (March 10, 1961), pp. 673–679.

MARKHAM, S. F. *Climate and the Energy of Nations*. New York: Oxford University Press, 1947.

METCALF, R. D. *Visual Recovery Times from High Intensity Flashes of Light*, Report No. TR 58-232, Aero Medical Laboratory, Wright Air Development Center, Wright-Patterson Air Force Base, Dayton, Ohio, October, 1958.

MILANKOVITCH, M. "Mathematische Klimalehre und Astronomische Theorie der Klimaschwankungen," in *Handbuch der Klimatologie*, eds. W. P. KOEPPEN and R. GEIGER, Band I, Teil A. Berlin: Verlag von Gebruder Borntraeger, 1930.

MILLER, H., RILEY, M. B., BONDURANT, S., and HIATT, E. P. "Duration of Tolerance to Positive *G*," abstracted in *J. Avia. Med.*, **29**, No. 3 (March, 1958), p. 243.

MILLER, S. L. "Production of Organic Compounds under Possible Primitive Earth Conditions," *J. Am. Chem. Soc.*, **77**, No. 9 (May 12, 1955), pp. 2351–2361.

NEWBURN, R. L., JR. "The Exploration of Mercury, the Asteroids, the Major Planets and Their Satellite Systems, and Pluto," in *Advances in Space Science and Technology*, ed. F. I. ORDWAY, Vol. 3. New York: Academic Press, Inc., 1961.

NORTON, E. F. *The Fight for Everest: 1924*. New York: Longmans, Green & Co., Inc., 1925, p. 334.

OPARIN, A. I. *The Origin of Life* (1st ed.). New York: The Macmillan Company, 1938.

———. *The Origin of Life on the Earth* (3rd ed.). New York: The Macmillan Company, 1957.

———. "The Origin of Life on the Earth," *New Scientist*, **11**, No. 249 (August 24, 1961), pp. 474–475.

OTIS, A. B., and BEMBOWER, W. C. "Effect of Gas Density on Resistance to Respiratory Gas Flow in Man," *J. Appl. Physiol.*, **2** (December, 1949), pp. 300–306.

PALM, C., and CALVIN, M. "Irradiation of Methane, Ammonia, Hydrogen and Water," *Bio-Organic Chemistry Quarterly Report*, University of California Radiation Laboratory Report No. UCRL-9519, January 31, 1959.

PUGH, G., and WARD, M. "Physiology and Medicine," Appendix VII in *The Conquest of Everest*, by J. HUNT. New York: E. P. Dutton & Co., Inc., 1954.

ROCHE, E. A. "Essai sur la constitution et l'origine du système solaire," *Acad. de Montpellier, Sciences*, **8** (1873), p. 235.

ROMAN, N. G. "Planets of Other Suns." Paper presented before the American Astronomical Society at Toronto, September 1, 1959.

RUSSELL, H. N. "Relations between the Spectra and Other Characteristics of the Stars," *Popular Astronomy* (1914), pp. 22, 275–294, 331–351.

SÄNGER, E. "Die Erreichbarkeit der Fixsterne," *Proceedings of the VII International Astronautical Congress, Rome, 17–22 September 1956*, Rome: Associazione Italiana Razzi, 1956, pp. 97–113. (*The Attainability of the Fixed Stars*, translated by R. Schamberg, RAND Corporation Translation No. T-69, The RAND Corporation, December, 1956.

———. "Nuclear Rockets for Space Flight," *Astronaut. Sci. Rev.*, **3**, No. 3 (July–September, 1961), pp. 9–15.

———. "Some Optical and Kinematical Effects in Interstellar Astronautics," *J. Brit. Interplanet. Soc.*, **18**, No. 7 (January–February, 1962), pp. 273–277.

SHAPLEY, H. *Of Stars and Men*. Boston: Beacon Press, 1959.

SIDGWICK, N. V. *The Chemical Elements and Their Compounds*, Vol. I. New York: Oxford University Press, 1950.

SIMONS, H. "New Evidence Found; Plant Life on Earth 2.7 Billion Yrs. Ago," *Los Angeles Times*, December 9, 1962.

SPECTOR, W. S. (ed.). *Handbook of Biological Data.* Wright Air Development Center, Wright-Patterson Air Force Base, Dayton, Ohio, October, 1956.
SPENCER, D. F., and JAFFE, L. D. *Feasibility of Interstellar Travel.* California Institute of Technology Jet Propulsion Laboratory Technical Report No. 32-233, March, 1962.
SPITZER, L., JR. "The Terrestrial Atmosphere above 300 Km," in *The Atmosphere of the Earth and Planets* (2nd ed.), ed. G. P. KUIPER. Chicago: University of Chicago Press, 1952, pp. 211–247.
STRUGHOLD, H. "The Ecosphere of the Sun," *Avia. Med.*, **26,** No. 4 (August, 1955), pp. 323–328.
TAX, S. (ed.). *The Evolution of Life.* (*Evolution after Darwin*, Vol. 1.) Chicago: University of Chicago Press, 1960.
TOBIAS, C. A., and SLATER, J. V. "Our View of Space Biology Widens," *Astronautics*, **7,** No. 1 (January, 1962), pp. 20–22, 47–52.
TROITSKAIA, O. V. "About the Possibility of Plant Life on Mars," *Astron. Zh.*, **19,** No. 1 (1952). Translated by J. L. ZYGIELBAUM, California Institute of Technology Jet Propulsion Laboratory, Pasadena, California, February 16, 1952.
UREY, H. C. *The Planets, Their Origin and Development.* New Haven: Yale University Press, 1952.
———. "The Atmospheres of the Planets," *Encyclopedia of Physics*, **52,** ed. S. FLÜGGE. Berlin: Springer-Verlag, 1959.
VAN DE KAMP, P. "Visual Binaries," *Encyclopedia of Physics*, **50,** ed. S. FLÜGGE. Berlin: Springer-Verlag, 1958.
VAN DEN BOS, W. H. "The Visual Binaries," in *Vistas in Astronomy*, Vol. II, ed. ARTHUR BEER. New York: Pergamon Press, Inc., 1956, pp. 1035–1039.
VON WEIZSÄCKER, C. F. "Über die Entstehung des Planetensystems" (Regarding the Origin of the Planetary System), *Z. Astrophys.*, **22** (1944), pp. 319–355.
WALD, G. "The Origin of Life," *Sci. Am.*, **191,** No. 2 (August, 1954), pp. 44–53.
———. "Life and Light," *Sci. Am.*, **201,** No. 4 (October, 1959), pp. 92–108.
WEBSTER, A. G. *The Dynamics of Particles and of Rigid, Elastic and Fluid Bodies.* Leipzig: B. G. Teubner, 1925.
WEISS, F. J. "The Useful Algae," *Sci. Am.*, **187,** No. 6 (December, 1952), pp. 15–17.
WILSON, C. L. *Production of Gas in Human Tissues at Low Pressures.* School of Aviation Medicine Report No. 61-105, Brooks Air Force Base, San Antonio, Texas, August, 1961.
WOYTINSKY, W. S., and WOYTINSKY, E. S. *World Population and Production, Trends and Outlook.* New York: The Twentieth Century Fund, 1953.
WULFECK, J. W., WEISZ, A., and RABEN, M. W. *Vision in Military Medicine.* Wright Air Development Center Technical Report No. 58-339, Wright Air Development Center, Wright-Patterson Air Force Base, Dayton, Ohio, November, 1958.
WUNDER, C. C. "Food Consumption of Mice during Continual Centrifugation," *Iowa Acad. Sci. Publ.*, **68** (1961), p. 616.
WUNDER, C. C., LUTHERER, L. O., and DODGE, C. H. "Survival and Growth of Organisms during Life-long Exposure to High Gravity." Lecture before the Aerospace Medical Association, Atlantic City, New Jersey, April 10, 1962.

Index

The letter *t* preceding a page number indicates tabular material.

Manufactured in the United States of America

Selected RAND Books

AKHMANOVA, O. S., FRUMKINA, R. M., MEL'CHUK, I. A., and PADUCHEVA, E. V. *Exact Methods in Linguistic Research*, translated by D. G. Hays and D. V. Mohr. Berkeley and Los Angeles: University of California Press, 1963.

BAKER, C. L., and GRUENBERGER, F. J. *The First Six Million Prime Numbers.* Madison: The Microcard Foundation, 1959.

BELLMAN, R. *Dynamic Programming.* Princeton: Princeton University Press, 1957.

BELLMAN, R (ed.). *Mathematical Optimization Techniques.* Berkeley and Los Angeles: University of California Press, 1963.

BRODIE, B. *Strategy in the Missile Age.* Princeton: Princeton University Press, 1959.

BUCHHEIM, R. W., and the Staff of The RAND Corporation. *Space Handbook: Astronautics and Its Applications.* New York: Random House, Inc., 1959.

DANTZIG, G. B. *Linear Programming and Extensions.* Princeton: Princeton University Press, 1963.

DUBYAGO, A. D. *The Determination of Orbits*, translated from the Russian by R. D. Burke, G. Gordon, L. N. Rowell, and F. T. Smith. New York: The Macmillan Company, 1961.

EDELEN, D. G. B. *The Structure of Field Space: An Axiomatic Formulation of Field Physics.* Berkeley and Los Angeles: University of California Press, 1962.

GRUENBERGER, F. J., and MCCRACKEN, D. D. *Introduction to Electronic Computers.* New York: John Wiley & Sons, Inc., 1963.

HALPERN, M. *The Politics of Social Change in the Middle East and North Africa.* Princeton: Princeton University Press, 1963.

HARRIS, T. E. *The Theory of Branching Processes.* Berlin: Springer-Verlag, 1963.

HEARLE, E. F. R., and MASON, R. J. *A Data Processing System for State and Local Governments.* Englewood Cliffs: Prentice-Hall, Inc., 1963.

HIRSHLEIFER, JACK, DEHAVEN, J. C., and MILLIMAN, J. W. *Water Supply: Economics, Technology, and Policy*. Chicago: The University of Chicago Press, 1960.

HITCH, C. J., and MCKEAN, R. *The Economics of Defense in the Nuclear Age*. Cambridge: Harvard University Press, 1960.

JOHNSON, J. J. (ed.). *The Role of the Military in Underdeveloped Countries*. Princeton: Princeton University Press, 1962.

KECSKEMETI, P. *Strategic Surrender: The Politics of Victory and Defeat*. Stanford: Standford University Press, 1958.

KERSHAW, J. A., and MCKEAN, R. N. *Teacher Shortages and Salary Schedules*. New York: McGraw-Hill Book Company, Inc., 1962.

KRIEGER, F. J. *Behind the Sputniks: A Survey of Soviet Space Science*. Washington, D.C.: Public Affairs Press, 1958.

LUBELL, H. *Middle East Oil Crises and Western Europe's Energy Supplies*. Baltimore: The Johns Hopkins Press, 1963.

MARKOWITZ, H. M., HAUSNER, B., and KARR, H. W. *SIMSCRIPT: A Simulation Programming Language*. Englewood Cliffs: Prentice-Hall, Inc., 1963.

MCKEAN, R. N. *Efficiency in Government through Systems Analysis: With Emphasis on Water Resource Development*. New York: John Wiley & Sons, Inc., 1958.

NEWELL, A. (ed.). *Information Processing Language-V Manual*. Englewood Cliffs: Prentice-Hall, Inc., 1961.

O'SULLIVAN, J. J. (ed.). *Protective Construction in a Nuclear Age*. 2 vols. New York: The Macmillan Company, 1961.

SOKOLOVSKII, V. D. *Soviet Military Strategy*, translated and annotated by H. S. Dinerstein, L. Gouré, and T. W. Wolfe. Englewood Cliffs: Prentice-Hall, Inc., 1963.

TANHAM, G. K. *Communist Revolutionary Warfare: The Viet Minh in Indochina*. New York: Frederick A. Praeger Inc., 1961.

WHITING, A. S. *China Crosses the Yalu: The Decision To Enter the Korean War*. New York: The Macmillan Company, 1960.

WILLIAMS, J. D. *The Compleat Strategyst: Being a Primer on the Theory of Games of Strategy*. New York: McGraw-Hill Book Company, Inc., 1954.